Probability Models

AF691385

TEACHER'S EDITION

DATA-DRIVEN MATHEMATICS

Patrick Hopfensperger, Henry Kranendonk, and Richard L. Scheaffer

Dale Seymour Publications®
White Plains, New York

This material was produced as a part of the American Statistical Association's Project "A Data-Driven Curriculum Strand for High School" with funding through the National Science Foundation, Grant #MDR-9054648. Any opinions, findings, conclusions, or recommendations expressed in this publication are those of the authors and do not necessarily reflect the views of the National Science Foundation.

Managing Editor: Alan MacDonell

Senior Mathematics Editor: Carol Zacny

Project Editor: Nancy R. Anderson

Production/Manufacturing Director: Janet Yearian

Production Coordinator: Roxanne Knoll

Design Manager: Jeff Kelly

Text and Cover Design: Christy Butterfield

Cover Photo: Stephen Frisch

This book is published by Dale Seymour Publications®, an imprint of the Alternative Publishing Group of Addison Wesley Longman, Inc.

Dale Seymour Publications
10 Bank Street
White Plains, NY 10606
Customer Service: 800–872–1100

Printed in the United States of America.

Order number DS21180

ISBN 1-57232-241-1

1 2 3 4 5 6 7 8 9 10-ML-03 02 01 00 99 98

This Book Is Printed On Recycled Paper

DALE SEYMOUR PUBLICATIONS®

Authors

Patrick Hopfensperger
Homestead High School
Mequon, Wisconsin

Henry Kranendonk
Rufus King High
Milwaukee, Wisconsin

Richard Scheaffer
University of Florida
Gainesville, Florida

Consultants

Jack Burrill
National Center for Mathematics
Sciences Education
University of Wisconsin-Madison
Madison, Wisconsin

Emily Errthum
Homestead High School
Mequon, Wisconsin

Maria Mastromatteo
Brown Middle School
Ravenna, Ohio

Vince O'Connor
Milwaukee Public Schools
Milwaukee, Wisconsin

Jeffrey Witmer
Oberlin College
Oberlin, Ohio

Data-Driven Mathematics Leadership Team

Miriam Clifford
Nicolet High School
Glendale, Wisconsin

Kenneth Sherrick
Berlin High School
Berlin, Connecticut

Richard Scheaffer
University of Florida
Gainesville, Florida

James M. Landwehr
Bell Laboratories
Lucent Technologies
Murray Hill, New Jersey

Gail F. Burrill
National Center for Mathematics
Sciences Education
University of Wisconsin-Madison
Madison, Wisconsin

Acknowledgments

The authors thank the following people for their assistance during the preparation of this module:

- The many teachers who reviewed drafts and participated in field tests of the manuscripts
- The members of the *Data-Driven Mathematics* leadership team, the consultants, and the writers
- Kathryn Rowe and Wayne Jones for their help in organizing the field-test process and the Leadership Workshops

Table of Contents

About *Data-Driven Mathematics* vi

Using This Module vii

Unit I: Random Variables and Their Expected Values

Lesson 1: **Probability and Random Variables** 3

Lesson 2: **The Mean as an Expected Value** 11

Lesson 3: **Expected Value of a Function of a Random Variable** 18

Lesson 4: **The Standard Deviation as an Expected Value** 24

Assessment: **Lessons 1–4** 32

Unit II: Sampling Distributions of Means and Proportions

Lesson 5: **The Distribution of a Sample Mean** 39

Lesson 6: **The Normal Distribution** 49

Lesson 7: **The Distribution of a Sample Proportion** 60

Assessment: **Lessons 5–7** 69

Unit III: Two Useful Distributions

Lesson 8: **The Binomial Distribution** 75

Lesson 9: **The Geometric Distribution** 86

Assessment: **Lessons 8 and 9** 95

Teacher Resources

Quizzes 101

Solutions to Quizzes 108

Procedures for Using the TI-83 Graphing Calculator 113

About *Data-Driven Mathematics*

Historically, the purposes of secondary-school mathematics have been to provide students with opportunities to acquire the mathematical knowledge needed for daily life and effective citizenship, to prepare students for the workforce, and to prepare students for postsecondary education. In order to accomplish these purposes today, students must be able to analyze, interpret, and communicate information from data.

Data-Driven Mathematics is a series of modules meant to complement a mathematics curriculum in the process of reform. The modules offer materials that integrate data analysis with high-school mathematics courses. Using these materials helps teachers motivate, develop, and reinforce concepts taught in current texts. The materials incorporate the major concepts from data analysis to provide realistic situations for the development of mathematical knowledge and realistic opportunities for practice. The extensive use of real data provides opportunities for students to engage in meaningful mathematics. The use of real-world examples increases student motivation and provides opportunities to apply the mathematics taught in secondary school.

The project, funded by the National Science Foundation, included writing and field testing the modules, and holding conferences for teachers to introduce them to the materials and to seek their input on the form and direction of the modules. The modules are the result of a collaboration between statisticians and teachers who have agreed on the statistical concepts most important for students to know and the relationship of these concepts to the secondary mathematics curriculum.

A diagram of the modules and possible relationships to the curriculum is on the back cover of each Teacher's Edition.

Using This Module

Why the Content Is Important

Data analysis is concerned with studying the results of an investigation that has already taken place, with the hope of discovering some patterns in the data that might lead to new insights into the behavior of one or more variables. Probability is concerned with anticipating the future, with the hope of discovering models that might allow the prediction of outcomes not yet seen. Of course, we cannot predict with certainty and so possible outcomes are generally stated along with their chances of occurring.

There is a connection between data and probability since the probabilities used for anticipating future events often come from the analysis of past events. Thus, a survey that says 60% of drivers do not wear seat belts serves as the basis for calculating the probability distribution for the number of drivers, out of the next ten observed, who are not wearing seat belts.

There is also a connection between the key components of describing distributions of data and the key components of describing probability distributions. The mean of the data parallels the expected value of the probability distribution, but notice the change in language from something we see as a fact to something we merely anticipate. Variation in data and in probability distributions is often measured by the standard deviation, but the calculation becomes the expected value of a function of a random variable in the latter case. Shapes of data distributions and probability distributions are described by the same terms—symmetric and skewed: so the context of such descriptions must be made clear.

In this module, students will learn about the connections between data analysis and probability. The emphasis is on the development of basic concepts of probability distributions, as contrasted with probability from counting rules, and the use of standard models for these distributions. The value of having such standard models is that you need study only a few probability distributions in order to solve a wide variety of probability problems. Students will see that a lot of mileage is obtained from the normal, binomial and geometric models.

The skills required for working through this module are mainly those of beginning algebra, except that an infinite series is introduced in Lesson 9. Experience with simulation would be helpful, as many ideas are introduced with this approach.

When teaching this module, it is important to emphasize the distinction between analyzing data to describe the past and building probability models for anticipating the future. We analyze data to discover what happened in the last Gallup poll. We use probability models before the poll is actually conducted to describe what might happen in the next one.

The language of probability requires that a chance event underlie the issue or phenomenon being studied. In this module, chance is thought of in terms of relative frequency. If an unbalanced die has probability 0.4 of coming up a 6, the implication is that after many tosses of the die approximately 40% of the outcomes would be a 6. This may sound obvious, but such a definition of probability rules out the discussion of such issues as "the probability of life on Mars" or "the probability that I will pass this test today."

Probability is sometimes difficult and subtle. It is also useful and fun. If you enter this study in the spirit of data analysis and investigation by simulation, it should be a rewarding educational experience for both you and your students.

Algebraic Content

- Variables and random variables
- Functions of random variables
- Summation notation
- Binomial coefficients
- Mathematical models

Statistics Content

- Probability distributions
- Expected values
- Standard deviation of a probability distribution
- Normal distribution as a model
- Properties of means and proportions through sampling distributions
- Applications of binomial distribution
- Applications of geometric distribution

Instructional Model

Probability Models is designed for student involvement and interaction. Each lesson opens with a statement of objectives, followed by some key questions of the type to be addressed in that lesson. The opening scenario is written to foster student discussion and serves to set the stage for the investigation to follow. Investigations almost always involve hands-on activities for the students, and students must work through these to get the full impact of the lesson. Investigations are followed by practice exercises that review and extend the material. Each lesson closes with a summary of the main points.

Working in small groups, students should attempt to answer all of the questions in the discussion sections and work through all of the practice exercises. Each unit is to be completed as a whole, since any one discussion item or problem is likely to build on what went before and may contain new information essential to what follows. This is not a unit in which you can assign the odd numbered exercises; all are important to the understanding of the material.

Although lessons should not be skipped and sections within lessons should not be skipped, it is not necessary to cover the whole module. Unit I covers the basics of random variables and probability distributions, including expected value. Unit II builds on the Unit I material and covers the normal distribution and how it is used as a model for evaluating properties of sample means and proportions. You could stop at this point and still have covered a useful and fairly complete set of lessons on probability distributions as models of reality. Unit III, covering the binomial and geometric distributions, contains lessons that are more specialized and more mathematical.

Whether two or three of the units are covered, the lessons should come close together, as each one in succession tends to build on the one that came before. It is not a good idea to use one of these lessons each Friday on the term, for example. Too much is forgotten in the interim.

Teacher Resources

At the back of this Teacher's Edition you will find:

- A quiz for each unit
- Solution keys for the quizzes
- Procedures for Using the TI-83 Calculator

Where to Use the Module in the Curriculum

A chapter on probability is usually found somewhere in the algebra sequence, but the material on probability in algebra books is often much abbreviated and weak in modern applications. The two probability modules in *Data-Driven Mathematics,* of which Probability Models is the second, can be used as replacements or supplements for these chapters. Since this material on random variables, probability distributions, and expected values requires subtle reasoning that is not common to most students, it is recommended that this module be used in the second algebra course, or al least no earlier than late in the first algebra course. This is more for the maturity of reasoning required than for any particular set of algebraic skills needed to do the work.

As to requirements, students should be familiar with the idea of a function, especially linear functions of the form $ax + b$, and should have some facility with using symbols. They should also have some experience with basic probability as a relative frequency. Before covering Lesson 8, make sure students have seen binomial expansions.

Technology

The calculations required in this module are not as extensive as would be required, for example, in a data analysis module. Nevertheless, students should have at least a graphing calculator available to them. Expected values and related quantities are readily calculated by such a calculator or a computer. Either can also handle the simulations required in this module.

Pacing/Planning Guide

The table below provides a possible sequence and pacing of the lessons.

LESSON	OBJECTIVES	PACING
Unit I: Random Variables and Their Expected Values		
Lesson 1: Probability and Random Variables	Understand the relative frequency concept of probability. Define random variables.	1 class period
Lesson 2: The Mean as an Expected Value	Understand how to compute and interpret the mean of a probability distribution.	1 class period
Lesson 3: Expected Value of a Function of a Random Variable	Find expected values of certain functions of random variables. Understand fair games.	1 class period
Lesson 4: The Standard Deviation as an Expected Value	Understand how to compute and interpret the standard deviation of data and probability distributions.	1–2 class periods
Unit II: Sampling Distributions of Means and Proportions		
Lesson 5: The Distribution of a Sample Mean	Understand the behavior of the distribution of means from random samples.	1 or more class periods depending upon experience with simulation
Lesson 6: The Normal Distribution	Understand the basic properties of the normal distribution. See the usefulness of the normal distribution as a model for sampling distributions.	1 class period
Lesson 7: The Distribution of a Sample Proportion	Gain experience working with proportions as summaries of data. Develop sampling distributions for sample proportions. Discover the meaning of margin of error in surveys.	1–2 class periods
Unit III: Two Useful Distributions		
Lesson 8: The Binomial Distribution	Understand the basic properties of the binomial distribution. Use the binomial distribution as a model for certain types of counts.	2 class periods
Lesson 9: The Geometric Distribution	Understand the basic properties of the geometric distribution. Use the geometric distribution as a model for certain types of counts.	1–2 class periods

Random Variables and Their Expected Values

LESSON 1

Probability and Random Variables

Materials: none
Technology: graphing calculators or computers (optional)
Pacing: 1 class period with extra time for homework

Overview

The lesson begins by looking at the distribution of the number of children per family, which is a variable taking on a finite number of integer values—discrete variables. The relative frequencies of the possible values become the estimated probabilities for those values as the lesson moves from data description to probability models. This connection, along with the distinction between the relative frequencies of data and the probabilities associated with a random variable, are the key parts of the lesson. The addition of probabilities for mutually exclusive events and complements are two basic probability concepts that are used with little introduction or explanation. These concepts may have to be reviewed.

Teaching Notes

Have students work through the investigations and practice exercises in small groups, if possible, with little instruction. In the discussions, make sure students understand the concepts of random variable—as distinguished from an ordinary variable taking on numerical values—and probability distribution. Also be sure that they understand what is meant by a model.

Follow-Up

Have students look for other distributions of data for variables of this discrete type. Have them display and plot the data, and describe a scenario in which the data distribution could serve as an estimate of a probability distribution.

STUDENT PAGE 3

LESSON 1

Probability and Random Variables

How many children are in a typical American family?

What is the probability of randomly choosing a family with two children?

What is a random variable?

According to the U.S. Bureau of the Census, the number of children under 18 years of age per family has a distribution as given on the table below. A "family" is defined as a group of two or more persons related by birth, marriage, or adoption, residing together in a household. In which category does your family belong?

OBJECTIVES

Understand the relative-frequency concept of probability.

Define random variables.

INVESTIGATE

Family Size

In reality, some families have more than four children under the age of 18. However, the number of such families is very small and their percent would be very small compared to the percents in this table. Thus, we can describe the essential features of the number of children per family by using this simplified table as a *model* of reality.

Number of Children	Percent of Families
0	51
1	20
2	19
3	7
4	3

PROBABILITY AND RANDOM VARIABLES 3

Solution Key

Discussion and Practice

1. **a.** 20%

b. 100 – 51 = 49%

c. Seven percent of families in the U.S. have exactly 3 children under the age of 18.

d. The percents add to 100, which is appropriate if all categories of children per family are accounted for. The paragraph that follows explains that this is an approximation.

2. **a.** The height of the bar is about 19 units, which represents the 19% of families that have exactly 2 children under the age of 18.

b. 51 + 20 + 19 = 90%

c. The distribution of number of children per family has a high point at 0 and then decreases through 4. The percent of families with more than 4 children is so small as to not show up on the table or graph. The distribution can be called skewed toward the greater numbers of children, that is, to the right.

STUDENT PAGE 4

Discussion and Practice

1. The first line of data in the table above is interpreted to mean that 51% of the families in the United States have no children under the age of 18.

a. What percent of families have one child under the age of 18?

b. What percent of families have at least one child under the age of 18?

c. How would you interpret the 7% for the "3 children" category?

d. What is the sum of the percents in the table? Explain why this is an appropriate value for the sum.

2. A bar graph of the data on the number of children per family is shown below.

a. What does the height of the bar over the 2 represent?

b. What percent of families have at most two children under the age of 18?

c. Describe in words the distribution of children per family.

Families with Children Under Age 18 per Family in the U.S.

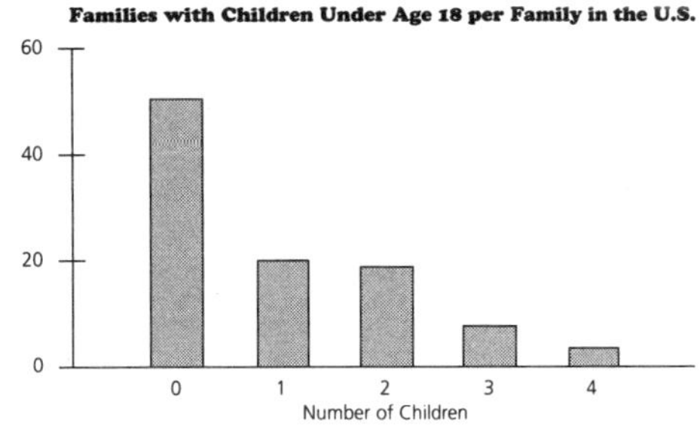

Random Variables

The discussion above makes use of the data table to describe one aspect of families in the United States. Suppose A. C. Nielsen, the company that provides ratings of TV shows, is planning to select a random sample of families from across the country. In that case, these same percents can be used as probabilities so that Nielsen can anticipate how many children under the age of 18 they might encounter in the sample.

4 LESSON 1

3. **a.** 0.20

b. 1 – 0.51 = 0.49

c. 0.51 + 0.20 + 0.19 = 0.90

d. 0.19 + 0.07 = 0.26

e. Based on the information given, the probability that a family has 5 children is 0.00. This does not imply that a family cannot have 5 children under the age of 18. The probability of such an event occurring could be, say, 0.002, which rounds to 0.00. Remember that the data in the table are only approximations to the true probabilities.

4. **a.** $P(C = 1)$

b. $P(C \geq 1)$

c. $P(C \leq 2)$

d. $P[(C = 2) \text{ or } (C = 3)] = P(C = 2) + P(C = 3)$

e. $P(C = 5)$

5. **a.** 0.51

b. 0.51 + 0.20 + 0.19 = 0.90

c. 0.07 + 0.03 = 0.10 = 1 – 0.90

d. 0.20 + 0.19 + 0.07 = 0.46

6. **a.** The probability that a randomly selected family has either 1 or 3 children under the age of 18

b. The probability that a randomly selected family has 2 or more children under the age of 18

STUDENT PAGE 5

3. Suppose Nielsen is to select one family at random. What is the approximate probability that the selected family will have

a. exactly one child under the age of 18?

b. at least one child under the age of 18?

c. at most two children under the age of 18?

d. either two or three children under the age of 18?

e. exactly five children under the age of 18?

In your past work, symbols have helped you to communicate mathematical statements more clearly and more concisely. Symbols can also help to clarify probability statements. In the situation above, the numerical outcome of interest is "the number of children under the age of 18 in a randomly selected U.S. family." Instead of writing this long statement each time we need it, why not just call it C? Then, C = the number of children under the age of 18 in a randomly selected U.S. family. From the data table, you can see that the probability that C is equal to 1 is 0.20, or 20%. It is cumbersome to write this probability statement in words, so we use a shorthand notation for the statement. The symbolic statement is $P(C = 1) = 0.20$.

When probability statements involve intervals of values for C, the symbolic form makes use of inequalities. For example, the probability that a randomly selected family has "at most one child under the age of 18" implies that the family has "either 0 or 1 child under the age of 18." This can be written as

$$P(C = 0) + P(C = 1) = P(0 \leq C \leq 1) = 0.51 + 0.20 = 0.71.$$

4. Write a symbolic statement for each of the statements in Problem 3.

5. Use the data table on page 3 to find the following probabilities.

a. $P(C = 0)$

b. $P(C \leq 2) = P[(C = 0) \text{ or } (C = 1) \text{ or } (C = 2)]$

c. $P(C \geq 3)$

d. $P(1 \leq C \leq 3)$

6. Write each of the following symbolic statements in words.

a. $P[(C = 1) \text{ or } (C = 3)]$

b. $P(C \geq 2)$

c. The probability that a randomly selected family has between 2 and 4 (2, 3, or 4) children under the age of 18

d. The probability that a randomly selected family has either 2 or fewer or 4 or more children under the age of 18 *or* The probability that a randomly selected family does not have exactly 3 children under the age of 18

7. **a.** $P[(C = 0) \text{ or } (C = 2) \text{ or } (C = 4)] = 0.51 + 0.19 + 0.03 = 0.73$

b. $P(C < 2) = P(C \leq 1) = 0.51 + 0.20 = 0.71$

c. $P(C \leq 1) = 0.51 + 0.20 = 0.71$

d. $P(C = 3) = 0 .07$

8. $P(C \leq 3) = 1 - 0.03 = 0.97$

STUDENT PAGE 6

c. $P(2 \leq C \leq 4)$

d. $P[(C \leq 2) \text{ or } (C \geq 4)]$

7. The ***complement*** of an event includes all possible outcomes except the ones in that event. For each of the symbolic statements in Problem 6, write a symbolic statement for the *complement of the event* in question. Find the probability of each complement.

8. Write "the probability that there are no more than three children in a randomly selected family" in symbolic form and find a numerical answer for this probability.

The symbols like C used to represent numerical outcomes from chance processes are called ***random variables***. Random variables are the basic building blocks for working with probability in scientific investigations. Probability ***distributions*** for random variables can be conveniently displayed in a two-column table like the one shown below for the random variable C, the "number of children under 18 in a randomly selected family."

C	P(C)
0	0.51
1	0.20
2	0.19
3	0.07
4	0.03

The probability distribution for a random variable can also be displayed in a bar graph, like the one shown below for the random variable C.

Probability Distribution for the Random Variable C

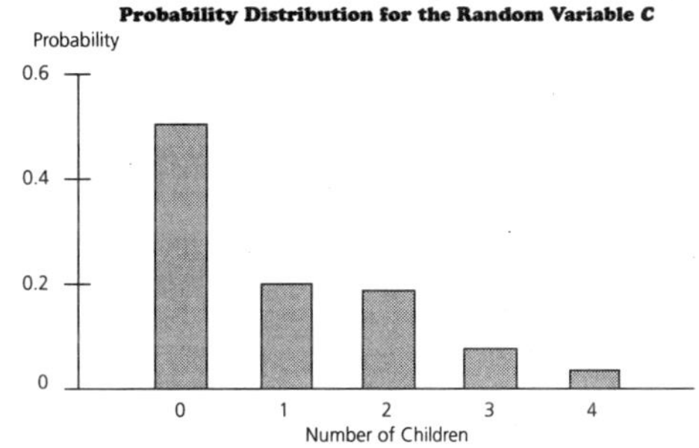

6 LESSON 1

9. a. Possible answer: The probability distribution of number of children per family has maximum at 0 and then decreases through 4. The probability of a randomly selected family having more than 4 children is so small as to not show up on the table or graph. The probability distribution can be called skewed toward the greater numbers of children, that is, skewed to the right.

b. Possible answer: The figures in Problems 2 and 8 have the same shape. In Problem 2, the vertical scale is percent, while in Problem 8 it is probability, which is measured on the interval 0 to 1. The distribution in Problem 2 describes the Census data on number of children per family. The distribution in Problem 8 describes the chances of getting a family with a certain number of children by random selection from among the families in the U.S.

c. The probabilities add to 1, as they account for a complete set of non-overlapping outcomes to the event "Select a family at random from the population of the U.S." It is possible to select a family with more than 5 children, but the probability would round to 0.00.

Practice and Applications

10. a. No. Possible answer: A household could have more than 4 cars, but the percent of such households would round to 0.0.

STUDENT PAGE 7

9. Study the relationship between the probability distribution as expressed in the table and as expressed in the graph.

a. Describe in words the shape of the probability distribution shown above.

b. What are the differences between the graphs in Problem 2 and Problem 8? Describe the different purposes they serve.

c. Add the column of probabilities in the table for the random variable C. What should be the sum of the probabilities in a complete probability distribution? Explain why this must be the case.

Practice and Applications

10. Consider another relative frequency distribution that can be turned into a probability distribution for a random variable. According to the *Statistical Abstract of the United States* (1996), the number of motor vehicles available to American households is given by the percents shown in the following table. A "household" is defined as all persons occupying a housing unit such as a house, an apartment, or a group of rooms. Notice the difference between a family and a household.

Number of Motor Vehicles per Household	Percent of Households
0	1.4
1	22.8
2	43.7
3	21.5
4	10.6

Note: Very few households have more than four motor vehicles.

The data in the table considers any transportation device that requires a motor-vehicle registration by the state in which it is located. For convenience, however, we will refer to these motor vehicles as "cars."

a. The sum of the percents in the table is 100. Does that mean that no household has more than four cars? Explain.

b. Define a random variable *Y* to be the number of cars available to a randomly selected American household. Construct the probability distribution for *Y* in table form.

PROBABILITY AND RANDOM VARIABLES 7

b.

Y	*P*(*Y*)
0	0.014
1	0.228
2	0.437
3	0.215
4	0.106

c.

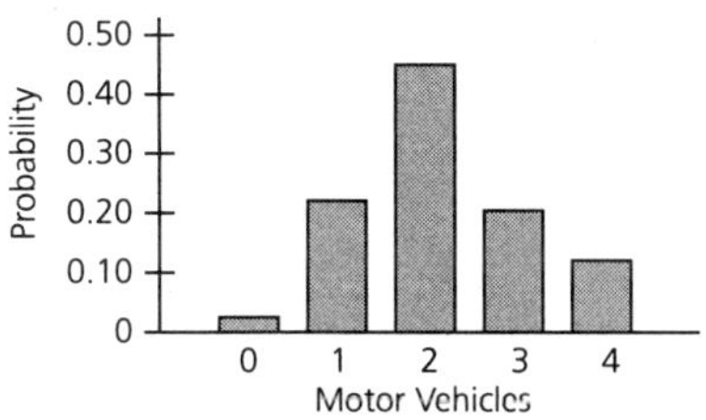

The probability distribution is mound-shaped and somewhat symmetric, with a higher probability at 4 than at 0.

11. **a.** 0.014

b. 0 .228

c. 1 – 0.014 = 0.986

d. 0.014 + 0.228 + 0.437 = 0.679

e. 0.215

12. **a.** $P(Y = 0)$

b. $P(Y = 1)$

c. $P(Y \geq 1) = 1 - P(Y = 0)$

d. $P(Y \leq 2)$

e. $P(Y = 3)$

13. **a.** 0.437; what is the probability that a randomly selected household has exactly two cars?

b. 1.0; what is the probability that a randomly selected household either has a car or does not have a car?

c. 1 – 0.106 = 0.894; what is the probability that a randomly selected household has at most three cars?

d. 0.228 + 0.437 + 0.215 = 0.880; what is the probability that a randomly selected household has one, two, or three cars?

14. $P(Y \geq 1) = 1 - P(Y = 0) = 1 - 0.014 = 0.986$

STUDENT PAGE 8

c. Construct a bar graph that represents the probability distribution for *Y*. Describe the shape of this distribution.

11. An automobile manufacturer is planning to conduct a survey on what Americans think about automobile repairs. What is the chance that a randomly selected household in the poll has

a. no cars?

b. exactly one car?

c. at least one car?

d. at most two cars?

e. exactly three cars?

12. Write each of the statements in Problem 11 in symbolic form.

13. With *Y* defined as in Problem 10, find each of the following probabilities and write the symbolic statement in words.

a. $P(Y = 2)$

b. $P(Y \geq 0)$

c. $P(Y \leq 3)$

d. $P(1 \leq Y \leq 3)$

14. For the first randomly selected household contacted, what is the probability that the household has at least one car? Write a symbolic statement for this probability.

15. The graph below provides information on how young adults are postponing marriage.

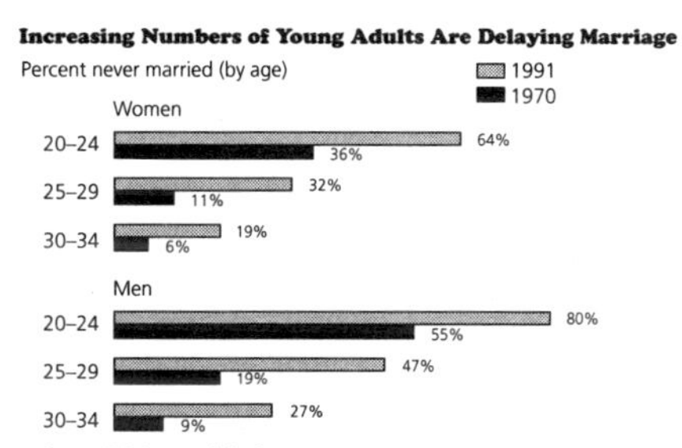

8 LESSON 1

STUDENT PAGE 9

15. a. Yes; for each age group, the percent of men who had never married is greater than the percent of women who had never married. This is true for both 1970 and 1991.

b. 1 – 0.64 = 0.36

c. 0.47

d. The data give only the percent never married among the males or females in a certain age class. No information is provided on the number of people in the various age classes. Therefore, we cannot estimate the probability that a randomly selected adult would be under age 34.

a. Do men tend to postpone marriage longer than women do? Use the data from the graph to support your answer.

b. Suppose a 1991 survey randomly sampled women between the ages of 20 and 24. What is the probability that the first such woman sampled was married?

c. Suppose a 1991 survey randomly sampled men between the ages of 25 and 29. What is the probability that the first such man sampled had never married?

d. From these data, can we answer the following question? Explain. "What is the probability that an adult randomly selected in a 1991 survey was under the age of 34 and had never married at the time of his or her selection?"

SUMMARY

A display, such as a table or a graph, showing the numerical values that a variable can take on and the percent of time that the variable takes on each value is called the *distribution* of that variable. If possible values of the variable are randomly selected, the variable is called a *random variable* and the percents attached to the numerical values give the probability distribution for that variable.

LESSON 2

The Mean as an Expected Value

Materials: none
Technology: graphing calculators (optional)
Pacing: 1 class period

Overview

The mean of a set of data arranged in a frequency table can be calculated by knowing either the frequencies or the relative frequencies for each data value. The relative frequencies may be displayed as a percent or as a fraction or decimal. When the relative frequencies are viewed as probabilities of future events, the calculation for the mean remains the same but the result is called the expected value of the random variable. A general formula for the expected value is developed using summation notation.

Teaching Notes

Allow students to work through the lesson in small groups. The summation notation used in the general formula for expected value will require some discussion, especially if it has not been seen before. Make sure students understand the subtleties of the difference between a mean of a data distribution and the expected value of a probability distribution.

Technology

Here, a graphing calculator is extremely useful for calculating expected values. For example, the TI-83 allows you to calculate the expected value in one step if the data values are in one list and the relative frequencies in another, using decimals. Use the STAT/CALC/1-Var Stats command with the data list named first and the relative frequency list second.

Follow-Up

Have students suggest other data sets or probability distributions for which the expected value would be a meaningful summary.

STUDENT PAGE 10

LESSON 2

The Mean as an Expected Value

What is the average number of children per family in America?

In a randomly chosen family, how many children would you expect to see?

How does the mean number of children per family compare to the mean family size?

OBJECTIVE

Understand how to compute and interpret the mean of a probability distribution.

An average, such as the arithmetic ***mean*** or simply the mean, is a common measure of the center of a set of data. The mean score of your quizzes in mathematics is, no doubt, an important part of your grade in the course. The mean age of residences in your neighborhood helps insurance companies figure out how much to charge for fire insurance. The mean amount paid by families for typical goods and services this year as compared to last year determines the rate of inflation. In this lesson, we will look at means of distributions of data to discover how they relate to means of probability distributions.

INVESTIGATE

How would you calculate the mean score of your quizzes in mathematics? In what other situations might you need to calculate the mean for a set of data?

Discussion and Practice

1. A football team played nine games this season, scoring 12 points in each of three games and 21 points in each of the other six games.

10 LESSON 2

Solution Key

Discussion and Practice

1. a.

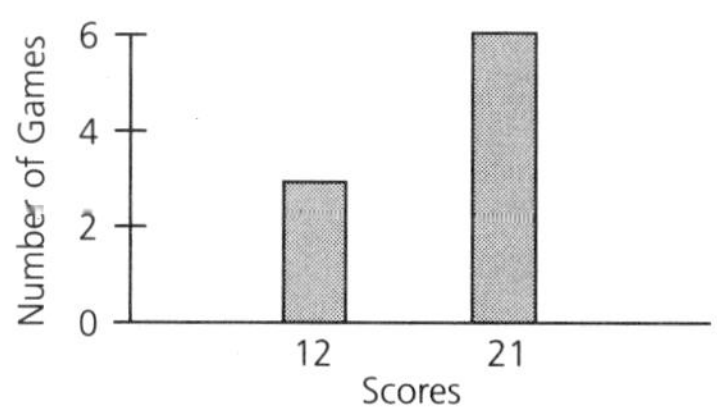

b. 18

c. The mean is closer to 21 since this score has the higher frequency and, hence, draws the mean toward it.

2. a.

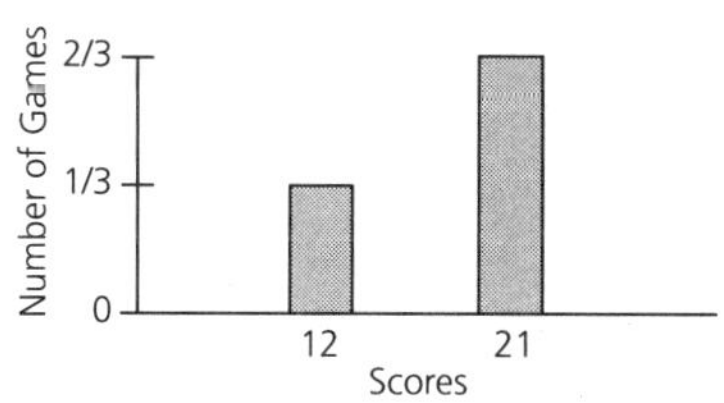

Possible answer: It is similar except that the vertical axis represents fraction of games played rather than number of games played.

b. $12(\frac{1}{3}) + 21(\frac{2}{3}) = 18$

c. Again, the mean is closer to the score of 21, which has the greater relative frequency.

3. The expected score for the future is the mean score from the past, 18.

4. Possible answer: The actual data are available for this year and the true mean score can be calculated. This year's record is used as a probability distribution for next year's record. In this sense, it tells us what is anticipated, or expected, to happen. The expected value is the mean score we anticipate, or expect, at this point, but it may not actually occur.

STUDENT PAGE 11

a. Construct a bar graph for the points scored, with the values for the variable "points scored" on the horizontal axis.

b. What is the mean number of points scored per game for this team? Explain how you found this mean.

c. Mark the value of the mean on the horizontal axis of the bar graph. Is the mean closer to 12 or to 21?

2. Suppose you knew that the team scored 12 points in $\frac{1}{3}$ of its games and 21 points in $\frac{2}{3}$ of its games, but you were not told how many games the team played.

a. Construct a bar graph for these data. How does it compare to the one in Problem 1a?

b. Can you still calculate the mean number of points per game? If so, what is it? Discuss how you arrived at this answer.

c. Mark the mean on the horizontal axis of the bar graph.

3. The team is expected to perform next year about as well as it performed this year. That is, the probability of scoring 12 points in a game is about $\frac{1}{3}$, while the probability of scoring 21 points in a game is about $\frac{2}{3}$. For a randomly selected game, how many points would you expect the team to score?

4. A mean computed from a probability distribution—an anticipated distribution of outcomes—is called an ***expected value***. Discuss why you think this terminology is used. Does the terminology seem appropriate?

5. Instead of a randomly selected game from next year's schedule, suppose we consider the game against the best team in the league. Would that change your opinion on the team's expected number of points scored? Why or why not?

Expected Value

Recall that one of the first numerical summaries of a set of data that you studied was the mean, used as a measure of center. We now review the calculation of the mean by working through an example. A survey of a class of 20 students reveals that 4 have no pets, 10 have one pet, and 6 have two pets. The data are shown in the table below.

THE MEAN AS AN EXPECTED VALUE 11

5. Yes; the expected number of points scored against the best team should be a little less than the average number of points scored against all opponents.

6. $(\frac{22}{20}) = 1.1$; $0(\frac{4}{20}) + 1(\frac{10}{20}) + 2(\frac{6}{20}) = 1.1$

7. **a.** 20%, 50%, 30%

b. $0(0.2) + 1(0.5) + 2(0.3) = 1.1$

c. The answers are the same, based on either frequencies or relative frequencies.

8. **a.** 0.91

b. The mean is not in the center of the graph. It is much closer to 0 because of the high frequency there.

STUDENT PAGE 12

Number of Pets	Number of Students	Total Number of Pets
0	4	0
1	10	10
2	6	12

The mean number of pets per student can be calculated in a variety of ways.

6. What is the mean number of pets per student? Discuss your calculation method with someone else in the class who used a different method.

7. Suppose we do not know how many students were surveyed, but we do know the percent of students who had each number of pets.

a. What percent of the students have no pets? One pet? Two pets? Add a column to the table for these percents.

b. Based on the percents in part a, find the mean number of pets per student surveyed. Explain how you arrived at your answer.

c. How does the answer compare to your answer for Problem 6? Should the answers be the same?

8. We now return to the Lesson 1 data on the number of children under the age of 18 per U.S. family.

Number of Children	Percent of Families
0	51
1	20
2	19
3	7
4	3

a. Use what you just learned to calculate the mean number of children per family in the U.S.

b. Find the mean on the horizontal scale of the bar graph for these data provided in the following graph. Is the mean in the center of the distribution? Why or why not?

12 LESSON 2

9. $E(C) = 0.91$

10. Problem 8 asks about the actual mean number of children in the current population of families. Problem 9 asks about the anticipated mean number of children in a random sample of families yet to be selected.

11. a. The expected value is the average value of the number of children per family that we anticipate seeing after many families are selected in a sample. The average number of children per family, across many families, need not be an integer.

b. $100(0.91) = 91$

c. $2500(0.91) = 2275$

d. $4000 = 0.91n$;
$n = 4000/0.91 \approx 4396$

STUDENT PAGE 13

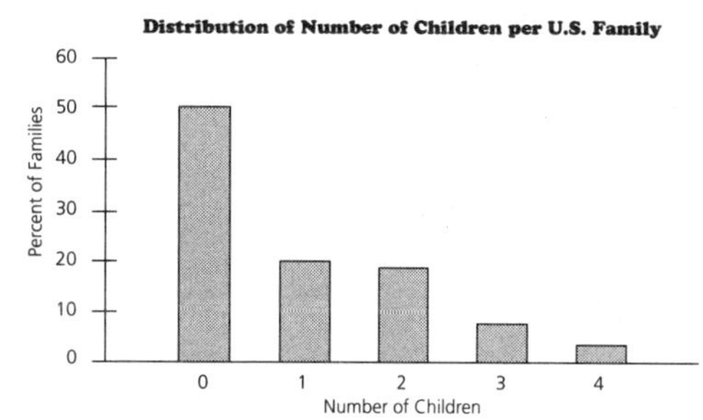

9. The A. C. Nielsen Company randomly selects families for use in estimating the ratings of TV shows. For each randomly selected family, how many children would we expect to see? That is, what is the expected value of C, the number of children in a randomly selected family? Show how you found your answer.

10. Explain why the terminology changed from *mean* number of children per family in Problem 8 to *expected* number of children per family in Problem 9.

11. The calculation of an expected value often results in a decimal. That is, the answer is not always an integer.

a. Explain why the decimal part of the expected number of children per family makes sense as an expected value, even though we cannot see a fraction of a child in any one family.

b. How many children in all would we expect to see in a random sample of 100 families?

c. How many children would we expect to see in a random sample of 2500 families?

d. If Nielsen really expects opinions from about 4000 children under the age of 18, how many families should be in the sample?

You now have the tools to develop a general expression for the expected value of a random variable.

12. Suppose a random variable X can take on values $x_1, x_2, \ldots, x_k$ with respective probabilities $p_1, p_2, \ldots, p_k$. That is, $P(X = x_i) = p_i$ for values of i ranging from 1 to k.

THE MEAN AS AN EXPECTED VALUE **13**

12. a. $E(X) = x_1p_1 + x_2p_2 + \ldots + x_kp_k$
The mean is found by multiplying each value of x by its relative frequency and adding the results. In a probability distribution, the expected value is found by the same procedure with the probabilities replacing the relative frequencies, since probability is a long-run relative frequency.

b. $E(X) = \sum_{i=1}^{k} x_i(p_i)$

Practice and Applications

13. a. $E(Y) = 2.58$, much greater than the expected number of children per family

b. $1000(2.58) = 2580$

c. $4000 = 2.58n$;
$n = 4000/2.58 \approx 1550$

STUDENT PAGE 14

a. Write a symbolic expression for the expected value of X. Explain the reasoning behind this expression.

b. A commonly used symbol for the expected value of X is $E(X)$, and $E(X)$ is expressed as a sum. The Σ symbol tells you to add the terms that follow the symbol, starting with the term indicated by the integer below the Σ and ending with the term indicated by the integer above the Σ. Thus,

$$\sum_{i=1}^{3} x_i p_i = x_1 p_1 + x_2 p_2 + x_3 p_3$$

Replace the question marks in the expression below with numerical values to indicate the range of summation.

$$E(X) = \sum_{i=?}^{?} x_i(?)$$

Practice and Applications

The following table shows the distribution of household sizes for U.S. households.

Number of Persons per Household	Percent of Households
1	25
2	32
3	17
4	16
5	7
6	2
7	1

Note: Households of more than 7 persons are very rare.

13. Suppose Nielsen is randomly sampling households in order to produce TV ratings. Let Y denote the size of a randomly selected household.

a. Find the expected value of Y. Compare it to the expected value of C found in Problem 9.

b. If Nielsen randomly selects 1000 households, how many people would these households be expected to contain?

c. If Nielsen really expects 4000 people in the survey, how many households should be sampled?

14 LESSON 2

14.

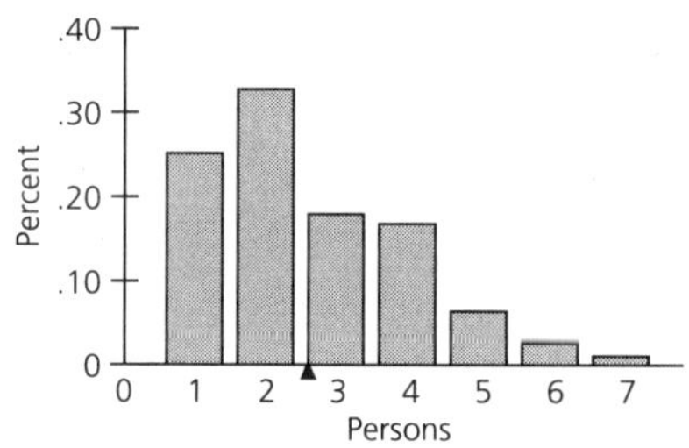

The expected value of 2.58 is to the left of center on the graph because the high probabilities at 1 and 2 pull the expected value in their direction.

STUDENT PAGE 15

14. Construct a bar graph of the probability distribution for *Y*. Mark the expected value of *Y* on the bar graph. Is the expected value in the center of the possible values for *Y*? Why or why not?

SUMMARY

The distribution of a variable is determined by the numerical values that the variable can take on, along with the proportion of times that each numerical value occurs. The ***mean*** of the distribution can be calculated from this information. If the proportions of times that the numerical values occur are interpreted as probabilities, then the mean is called the ***expected value*** of a probability distribution. The mean of a data set describes something that has already happened, while an expected value anticipates what might happen in the future.

THE MEAN AS AN EXPECTED VALUE **15**

LESSON 3

Expected Value as a Function of a Random Variable

Materials: none
Technology: scientific or graphing calculators (optional)
Pacing: 1 class period

Overview

The two goals of this lesson are to demonstrate that the expected value of a linear function is the same linear function of the expected value and to understand what is meant by a fair game. Payoffs for games often can be written as linear functions of random variables, and so the two topics naturally fit together.

Teaching Notes

As in previous lessons, these exercises are appropriate as small-group activities. You should direct some class discussion around the issue of fair games and how this idea applies to business decisions, for example.

Follow-Up

Have students construct some games of their own, perhaps using spinners or dice, for which they can calculate the probability of winning. Have them decide whether or not the games are fair.

STUDENT PAGE 16

LESSON 3

Expected Value of a Function of a Random Variable

How much would you expect to pay to feed a pet for a week?

What is a "fair" game?

How do business decisions depend on the concept of a fair game?

OBJECTIVES

Find expected values of certain functions of random variables.

Understand fair games.

As you have seen in previous work in mathematics and science, it is often convenient to express one variable as a function of another. Most goals in basketball are worth 2 points; hence, the number of points a player scores in a game, excluding free throws and 3-point goals, can be written as a function of the number of goals made.

Charlotte takes about 20 shots from inside the 2-point area during a game and expects to make about 60% of them. How many points can she expect to get from these goals?

INVESTIGATE

Expected Value of a Function

In practical applications of probability, the random variables are often written as functions of other random variables. Suppose the table below gives estimates of the probabilities of a randomly selected student having 0, 1, or 2 pets.

16 LESSON 3

Solution Key

Discussion and Practice

1. $E(X) = 0(0.2) + 1(0.5) + 2(0.3) = 1.1$

2. a.

Y	$P(Y)$
0	0.2
20	0.5
40	0.3

b. $E(Y) = 0(0.2) + 20(0.5) + 40(0.3) = 22$

3. a. $E(Y) = \sum_{i=1}^{k} y_i(p_i) = \sum_{i=1}^{k} 20x_i(p_i)$

$= 20 \sum_{i=1}^{k} x_i(p_i) = 20E(X)$ since the values for Y are simply 20 times the corresponding values for X.

b. $E(Y) = 20E(X) = 20(1.1) = 22$

4. $E(\text{Profit}) = E(\text{Gain}) - 20 = 15(8) - 20 = 100$

5. a. $E(Y) = \sum_{i=1}^{k} y_i(p_i)$

b. $\sum_{i=1}^{k} y_i(p_i) = \sum_{i=1}^{k} (ax_i + b)(p_i)$

c. $\sum_{i=1}^{k} (ax_i + b)(p_i)$

$= a\sum_{i=1}^{k} x_i(p_i) + b\sum_{i=1}^{k} (p_i)$

d. $a\sum_{i=1}^{k} x_i(p_i) + b\sum_{i=1}^{k} (p_i)$

$= aE(X) + b$

Note that the sum of the probabilities must be 1.

STUDENT PAGE 17

Number of Pets	Probability
0	0.2
1	0.5
2	0.3

If X represents the number of pets per randomly selected student, then $E(X)$ can be calculated from the information in the table.

Discussion and Practice

1. Calculate $E(X)$ for the pet example.
2. Suppose that it costs around $20 per week to feed a pet. We now want to study the probability distribution for a new random variable Y, the cost of pet feeding per week.
   a. Write the probability distribution for Y in table form, based on the information provided above on number of pets per student.
   b. Use the results of Problem 2a to calculate $E(Y)$, the expected weekly pet feeding cost per student.
3. Another way to find the expected value of Y is to observe that $Y = 20X$.
   a. Use the formula for $E(Y)$ and $E(X)$ to show that $E(Y) = 20E(X)$.
   b. Calculate $E(Y)$ by using the result in Problem 3a. Compare this result with your answer to Problem 2b.
4. Sam has a job mowing lawns. He expects to mow 8 lawns per week. He charges $15 per lawn, but it costs him about $20 per week to keep the lawn mower in good repair and full of fuel. What is Sam's expected profit per week?
5. Show that, in general, if $Y = aX + b$ then $E(Y) = aE(X) + b$.
   a. Begin by writing the formula for $E(Y)$ as a summation, assuming Y can take on values $y_1, y_2, \ldots, y_k$, with respective probabilities $p_1, p_2, \ldots, p_k$.
   b. Substitute $y_i = ax_i + b$ inside the summation.
   c. Use the Distributive Property to write the terms inside the summation as a sum of two terms.
   d. Use the properties of sums to write the summation as a term involving $E(X)$.

6. Cost = 500 + 100Y;
E(Cost) = 500 + 100$E(Y)$
= 500 + 100(2.58)
= $758

7. Time = 10 + 30C;
E(Time) = 10 + 30$E(C)$
= 10 + 30(0.91)
= 37.30 minutes.

8. Possible answer: A game is fair if the expected loss for the player is zero. This implies that the expected gain for the person running the game is also zero.

9. You should be willing to pay $0.50. Then, your gain is −0.50 with probability $\frac{199}{200}$ and 99.50 with probability $\frac{1}{200}$, for an expected gain of $E(G) = -0.50(\frac{199}{200}) + 99.50(\frac{1}{200}) = 0$

STUDENT PAGE 18

Practice and Applications

6. Refer to Problem 13 in Lesson 2. Suppose the cost to the Nielsen Company for connecting a family to their system is a flat rate of $500 per household plus $100 for every family member in the household. How much should Nielsen expect to pay per family for connection charges?

7. Refer to Problems 8 and 9 in Lesson 2. The Gallup organization wants to sample children under the age of 18 and ask them about their attitudes toward school. It cannot sample children directly but it can sample families. It takes about 10 minutes to question the family about the status of their children and about 30 additional minutes for each interview conducted. How much time, on the average, should Gallup allow for each family sampled?

Fair Games

In the town raffle, a drawing is to take place for a radio worth about $100. Two hundred tickets will be sold for $1 each. The tickets are mixed in a drum and one ticket is randomly selected for the winning prize. If you buy one ticket, let's analyze what happens to G, the amount you gain.

There are two possible outcomes: you win or you lose. If you lose you have lost $1, which can be called a gain of −1. If you win, however, you gain $100 minus the $1 you paid to play, for a net gain of $99. So the probability distribution for G is as shown in the table.

G	$P(G)$
−1	199/200
99	1/200

By the rules of expected value, your expected gain is

$$E(G) = -1(\tfrac{199}{200}) + 99(\tfrac{1}{200}) = -\tfrac{1}{2}.$$

You can expect to lose a half dollar for every play of such a game. Would you call this a fair game?

8. Write a reasonable definition of a fair game.

9. What would you be willing to pay for a drawing like the one above to make the drawing fair in the sense of expected gain?

18 LESSON 3

STUDENT PAGE 19

10. If the game is fair for you as a player, do the people running the drawing make any money? Do you see a reason why most games are not fair?

11. There is another way to assess the expected gain for the game described above. Suppose we define W as the amount you win. Then, your gain can be written as a function of W.

a. Find the probability distribution of W for the game described at the beginning of this investigation, in which you pay \$1 to play.

b. Find the expected value of W.

c. Write the player's gain G as a function of W, with G and W as defined for the game above.

d. Use $E(W)$ to find $E(G)$. Does the answer agree with what we found earlier? Which method seems easier?

12. N tickets are sold for a drawing that will have one randomly selected winner. The payoff is an amount A. Each ticket sells for an amount C.

a. Find the probability distribution for the winnings of a player who buys one ticket.

b. Find the expected winnings for a player who buys one ticket.

c. How much should each ticket cost if this is to be a fair game?

d. Do the answers to Problems 12b and 12c seem reasonable? Explain.

13. An insurance company insures a car for \$20,000. The one-year premium paid for the insurance is denoted by r. The company has records on drivers and cars of the type insured here and estimates that they will sustain a total loss with probability 0.01 and a 50% loss with probability 0.05. All other partial losses are ignored.

a. Find the probability distribution for the amount the company pays out.

b. Find the company's expected gain if r = \$1000.

c. What should the company charge as a premium to make this a "fair game"? Can the company actually do this? Explain.

10. For a fair game, the expected gain is zero for both the player and the operator of the game. This implies that the operator would make no money in the long run and, hence, could not stay in business. Commercial games, therefore, cannot be completely fair to the player.

11. a.

W	$P(W)$
0	$\frac{199}{200}$
100	$\frac{1}{200}$

b. $E(W) = 0(199/200) + 100(1/200) = 0.50$

c. $G = W - 1.00$

d. $E(G) = E(W) - 1.00 = 0.50 - 1.00 = -0.50$, or a loss of \$0.50.

The answer agrees with that of the earlier discussion. W has a slightly simpler probability distribution, with an expected value that is easy to compute mentally. Writing G as a function of W may be the easier method.

Practice and Applications

12. a.

W	$P(W)$
0	$\frac{N-1}{N}$
A	$\frac{1}{N}$

b. $E(W) = 0(\frac{N-1}{N}) + A(\frac{1}{N}) = \frac{A}{N}$

c. $E(\text{Gain}) = E(W) - C = \frac{A}{N} - C$;

if $E(\text{Gain}) = 0$, then $C = \frac{A}{N}$.

d. The expected amount won is the total amount available divided equally among all the tickets sold. For a fair game, a player should be willing to pay the average amount to be won per ticket.

13. a. The amount the company pays out can be thought of as the amount "won" by the client in this "game." Denote this by W.

W	$P(W)$
0	0.94
10,000	0.05
20,000	0.01

$E(W) = 10{,}000(0.05) + 20{,}000(0.01) = \$700$

b. $E(\text{Gain for the company}) = r - E(W) = r - 700 = 1000 - 700 = \$300$

c. $E(\text{Gain for the company}) = r - 700 = 0$ implies $r = \$700$. The company has a certain cost of doing business and, therefore, cannot stay in business unless its expected gain from the policies it sells is positive.

STUDENT PAGE 20

SUMMARY

Many practical applications of probability involve finding expected values of functions of random variables. For linear functions of the form $Y = aX + b$ for constants a and b,

$$E(Y) = aE(X) + b.$$

LESSON 4

The Standard Deviation as an Expected Value

Materials: none
Technology: graphing calculators (optional)
Pacing: 1–2 class periods

Overview

The standard deviation is introduced as a measure of variability for data displayed in a relative-frequency table, with immediate extension to the use of standard deviation as a measure of variability for a probability distribution. The variance is introduced as the average of the squared deviations from the mean, and the standard deviation is then the square root of the variance. Students are asked to develop a general expression for the variance using summation notation.

Teaching Notes

As in previous lessons, this lesson is set up so that the exercises can be completed by small groups. Have students work through the spread-sheet approach to the calculation of variance and standard deviation so that they can see the steps and understand what standard deviation measures.

Whether the divisor of the sample standard deviation should be n or $(n - 1)$ is always a confusing issue. When calculating the standard deviation from relative-frequency data, the size of the data set n may be unknown. The calculation, then, as outlined here is equivalent to dividing by n, should n be known. That is the correct way to calculate the standard deviation when it is used as a measure of variability in a probability distribution.

The idea that an interval of one standard deviation to either side of the mean will contain a somewhat predictable amount of the probability distribution merits some discussion. For mound-shaped symmetric distributions, this will be about 70% of the distribution. For skewed distributions, it will be more than 70%.

Technology

A graphing calculator will be a great help with these calculations. The suggested spread- sheet calculations can be done in the lists of such a calculator. Also, the standard deviation can be calculated directly by placing the values of the variable in one list and the relative frequencies, or probabilities, as decimals in another. The TI-83 does the calculation through the STAT/CALC/1-Var Stats command. If your calculator has two forms of the standard deviation built in, make sure students look at the one that is equivalent to dividing by n.

Follow-Up

Have students calculate and interpret the standard deviation for other frequency or relative frequency distributions of data, or for other probability distributions.

Have students find and report on an article from research literature that makes use of standard deviation in drawing its conclusions.

STUDENT PAGE 21

LESSON 4

The Standard Deviation as an Expected Value

Do most households have about the same number of cars, or is there a great deal of variation from household to household?

Is the number of persons per family more variable than the number of children per family?

How can you measure variation in a probability distribution?

In data analysis, once we have a measure of center, it is important to develop a measure of ***variation***, or spread, of the data to either side of the center. One useful measure of variability is the standard deviation, a value you may have encountered in lessons on data analysis. We now develop that same measure of spread for probability distributions.

A ***deviation*** is the distance between an observed data point and the mean of the distribution of data. The average of the squared deviations has a special name, ***variance***. The square root of the variance is called the ***standard deviation***. The standard deviation has important practical uses in probability and statistics, some of which we will see in future lessons of this unit.

OBJECTIVE

Understand how to compute and interpret the standard deviation of data and probability distributions.

INVESTIGATE

Recall the data in Lesson 2 regarding the number of pets students have. The survey of 20 students revealed that 4 have no pets, 10 have one pet, and 6 have 2 pets. A tabular array for

Solution Key

Discussion and Practice

1. **a.** The data points here are 0, 1, and 2. Thus, the deviations from the mean of 1.1 are –1.1, –0.1, and 0.9, respectively; answers will vary.

b. The average of the deviations is $\frac{4(-1.1) + 10(-0.1) + 6(0.9)}{20} = 0.$

c. The squared deviations are 1.21, 0.01, and 0.81, respectively.

d. The average of the "squared deviations from the mean" is $\frac{4(1.21) + 10(0.01) + 6(0.81)}{20} = \frac{9.8}{20}$ $= 0.49.$

2. The standard deviation is the square root of the average of the squared deviations, or square root of the variance, 0.7 in this case; answers will vary.

3.

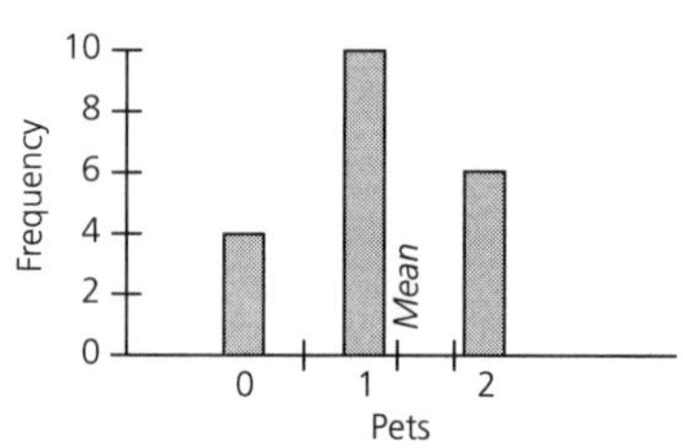

a. The mean is 1.1.

b. This point is at 1.1 + 0.7 = 1.8.

c. This point is at 1.1 – 0.7 = 0.4.

d. Only the data values of 1 are between these boundaries; these values comprise 50% of the 20 data values in the original data set.

STUDENT PAGE 22

these data follows. How could you calculate the numbers in the third column?

Number of Pets	Number of Students	Percent of Students
0	4	20
1	10	50
2	6	30

Discussion and Practice

The mean number of pets per student is 1.1. Do you remember how we determined this? The standard deviation is a special function of the variable "number of pets" and can be calculated by making use of what we learned in Lessons 2 and 3.

1. Use the following steps to find the variance of the number of pets per student.

a. Add a column of "deviations from the mean" to the table. What would you say is a "typical" deviation?

b. Find the average of the deviations from the mean.

c. Add a column of "squared deviations from the mean" to the table.

d. Find the average of the squared deviations from the mean, called the ***variance***.

2. The ***standard variation*** is the square root of the variance. Find the standard deviation of the number of pets per student. Is this number close to what you chose as a typical deviation in Problem 1a?

3. Draw a bar graph of the data on the number of pets per student given above.

a. Mark the mean of this distribution on the graph.

b. Mark off a distance of one standard deviation above, that is, to the right of, the mean.

c. Mark off a distance of one standard deviation below, that is, to the left of, the mean.

d. What fraction of the 20 data values are inside the interval from one standard deviation below the mean to one standard deviation above the mean?

4. Sketch another bar graph, still using data values of 0, 1, and 2 but choosing frequencies which would have greater

22 LESSON 4

4. Possible answer:

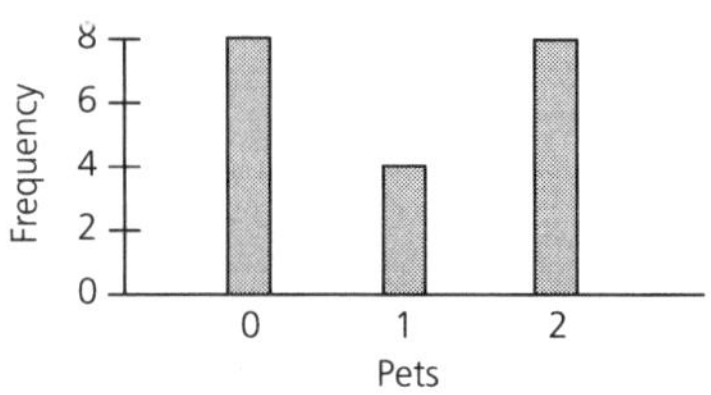

For this distribution, the mean is 1.0 and the standard deviation is 0.89, larger than the standard deviation of the distribution in Question 3. The standard deviation measures the spread of the data at both sides of the mean. This bar graph has greater spread than the one in Question 3 because of the higher frequencies at 0 and 2, with a corresponding lower frequency at 1.

5. **a.** 1.11

b. The mean is 0.91 children per family.

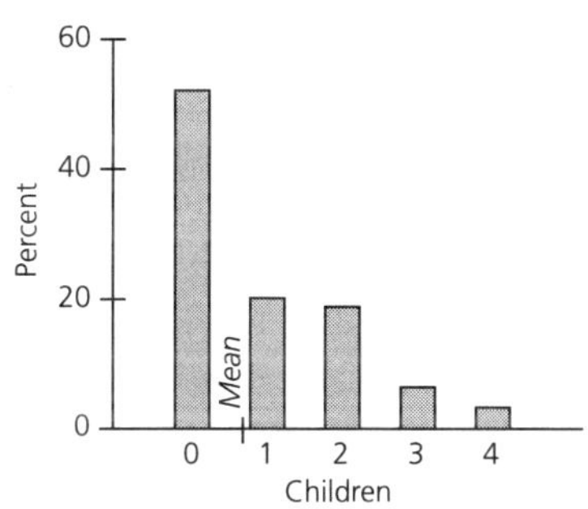

c. 0 0.91 – 1.11 = –0.20;
0.91 + 1.11 = 2.02

d. The data values 0, 1, and 2 are inside the interval marked off by the points of Problem 5c. These values comprise 90% of the possible values for number of children per family. Note that the lower boundary drops below zero and is outside the actual range of the data.

6. Possible answer: The standard deviation is a form of "average deviation from the mean." If we were to summarize the sizes of the deviations from the mean in a single number, this "average" is often a good summary value. An average can be thought of as a typical value in a data set.

7. **a.** $E(C) = 0.91$ and $SD(C) = 1.11$

b. The numerical values are the same.

c. In Problem 5, the mean and the standard deviation describe the center and spread of the *actual* number of children per family in the U.S. In Problem 7a, the expected value and the standard deviation describe the *anticipated* center and spread of values for the number of children per family in a sample that is not yet selected.

STUDENT PAGE 23

standard deviation than the one in Problem 3. Explain what feature of the bar graphs is measured by standard deviation.

Consider again the distribution of the number of children under the age of 18 in U.S. families as given in the table below.

Number of Children	Percent of Families
0	51
1	20
2	19
3	7
4	3

5. We now study this distribution using what you learned earlier in this lesson.

a. Calculate the standard deviation of the number of children per family. Recall that the mean was 0.91.

b. Sketch a bar graph of this distribution. Mark the mean number of children per family on the graph.

c. Mark off a distance of one standard deviation to both sides of the mean.

d. What percent of families would have a number of children inside the interval marked off in Problem 5c? How does this value compare with the answer to Problem 3d?

6. Sometimes the standard deviation is referred to as a "typical" deviation between a data point and the mean. Is this a fitting description? Explain.

7. The A. C. Nielsen Company plans to randomly select a large number of families to be used in collecting data for rating TV shows. Let *C* represent the random variable "number of children under the age of 18 in a randomly selected U.S. family."

a. Find the standard deviation we would expect for *C*, based on the available data and the fact that the expected value is 0.91.

b. Is there any difference between the numerical values for standard deviations calculated in Problems 5 and 7a?

c. Is there any difference in interpretation between the standard deviations calculated in Problems 5 and 7a? Explain.

THE STANDARD DEVIATION AS AN EXPECTED VALUE **23**

8. Using the symbol μ to denote $E(X)$,

$$\text{Variance}(X) = \sum_{1=1}^{n} (x_i - \mu)^2 p_i$$

and $$SD(X) = \sqrt{\sum_{1=1}^{n} (x_i - \mu)^2 p_i}.$$

NOTE: This is a big jump for students, as it moves from data to symbols. They may need some direction form the you at this point.

Discussion and Practice

9. **a.** E(Number of cars) = 2.17; SD(Number of cars) = 0.94

b. 2.17 – 0.94 = 1.23; 2.17 + 0.94 = 3.11

This interval covers the data values 2 and 3. About 65.2% of all households in the U.S. have a number of cars in this interval.

c. Possible answer: The average number of cars per household in the U.S. is 2.17. That implies that a sample of 100 typical households would have around 217 cars. The data on number of cars per household is concentrated from 1 to 3, which implies that most households have at least one car, but relatively few households have more than three cars.

STUDENT PAGE 24

You can now make the transition from working with numbers to working with symbols. The goal is to develop a formula for the standard deviation as calculated from a probability distribution.

8. Suppose a random variable X can take on the values $x_1, x_2, \ldots, x_n$ with respective probabilities $p_1, p_2, \ldots, p_n$. Write a symbolic expression for the standard deviation of X as an expected value.

Practice and Applications

9. Data on automobiles per family in the U.S. are given below.

Number of Cars	Percent of Households
0	1.4
1	22.8
2	43.7
3	21.5
4	10.6

a. Calculate the expected value and standard deviation of the number of cars per household that would be expected in a random sample of households from the U.S.

b. What percent of the households have a number of cars within one standard deviation of the mean?

c. Suppose you are allowed to use only the mean and standard deviation to describe these data in a newspaper article to be read by people who are not familiar with these terms. Write such a description.

10. The distribution of the number of persons per household in the U.S. is given in the following table.

Number of Persons per Household	Percent of Households
1	25
2	32
3	17
4	16
5	7
6	2
7	1

24 LESSON 4

STUDENT PAGE 25

10. a.

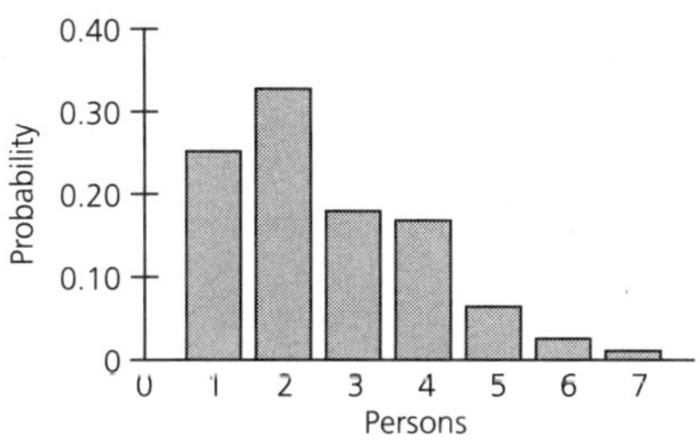

Possible answer: The distribution of the number of persons per household ranges from 1 to 7, with a concentration of values between 1 and 4; it is more spread out than the distribution of number of children per family. The latter has a concentration of values between 0 and 2. The standard deviation of the number of persons per household will be greater.

b. *SD*(Number of persons per household) = 1.39; *SD*(Number of children per family) = 1.11

11. Possible answer: If this additional information were available, the distributions would be more spread out, and the standard deviations would be greater.

12. Possible answer: For somewhat symmetric distributions, the interval mean ± 1*SD* includes the values concentrated in the middle of the distribution, and usually includes 50% to 70% of the possible data values. (See the distributions on number of pets and number of cars.) For highly skewed distributions, the interval mean ± 1*SD* includes the values concentrated at the end of the distribution with high frequencies, and usually includes more than 70% of the possible values, sometimes 90% or more. (See the distributions on number of children and number of persons.)

a. Sketch a bar graph for the distribution of the number of persons per household. Compare this distribution with the distribution of the number of children per family. Which will have the greater standard deviation? Explain why without calculating the standard deviation for the number of persons per household.

b. Calculate the standard deviation of the number of persons per household. Does it confirm your answer to Problem 10a?

11. In the tables showing the number of children per family and the number of people per household, the greatest value shown in the tables is not the greatest possible value. That is, there can be more than 4 children in a family and there can be more than 7 people in a household. If more accurate data on large families were available, what effect would that have on the calculated values of the standard deviations? Explain.

12. Looking at all the distributions seen so far in this lesson, for which does the standard deviation seem to be the best as a measure of a typical deviation from the mean? For which does it seem to be the worst? Explain.

13. The table below shows the percents of sports shoes of different types that are sold to various age groups.

Age of User	Gym Shoes	Jogging Shoes	Walking Shoes
Under 14	39.3	8.8	3.3
14 to 17	10.7	11.7	1.9
18 to 24	8.5	8.4	2.7
25 to 34	13.2	22.3	12.2
35 to 44	11.4	24.1	16.2
45 to 64	11.6	19.5	36.6
65 and over	5.3	5.2	27.1

Source: *Statistical Abstract of the United States*, 1993–94

a. Construct meaningful plots of the three age distributions. Comment on their differences. Which has the greatest mean? Which has the greatest standard deviation? You might begin by choosing a meaningful age to represent each of the shoe categories. Then, the data will look more like what we have been studying in this lesson and can be plotted as a bar graph.

b. Approximate the median age of user for each of the three shoe types.

THE STANDARD DEVIATION AS AN EXPECTED VALUE **25**

13. a. The underlying data, ages, is continuous rather than discrete; but it is difficult to construct a good histogram with the class summaries given here. One meaningful plot is simply a line graph across the midpoints of the age intervals. The "65 and over" class was arbitrarily cut off at 84 to make the last interval comparable to many of the others.

Possible answer:

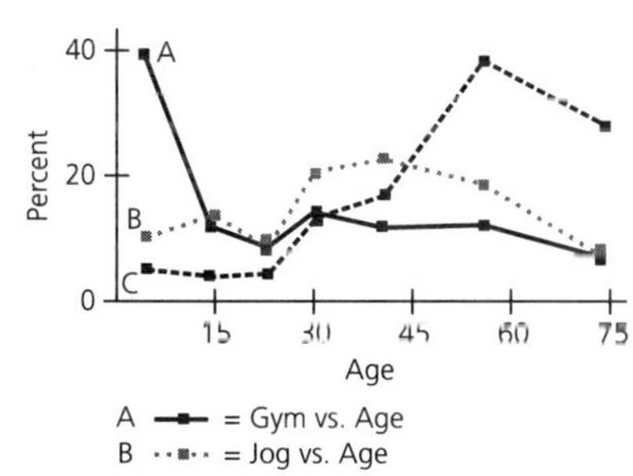

Gym shoes start out at high frequencies for the younger ages and then level off, whereas walking shoes start out low and end up with high frequencies at the older ages. Jogging shoes peak in the middle age groups.

Walking shoes must have the greatest mean age, whereas gym shoes must have the least.

It is very difficult to tell how the standard deviations compare by looking at the distributions. Since ages for jogging shoes have a peak in the middle, their standard deviation may be the least. Ages for gym shoes seem to have the flattest distribution, and that may give them the greatest standard deviation.

Another possible plot is the cumulative percent plot shown below. Here, the percents are added as we go up the age scale. For example, 50.0% of the gym-shoe buyers are age 17 or younger, and 83.1% are 44 or younger. This plot shows clearly that gym shoes have the youngest buyers and walking shoes the oldest.

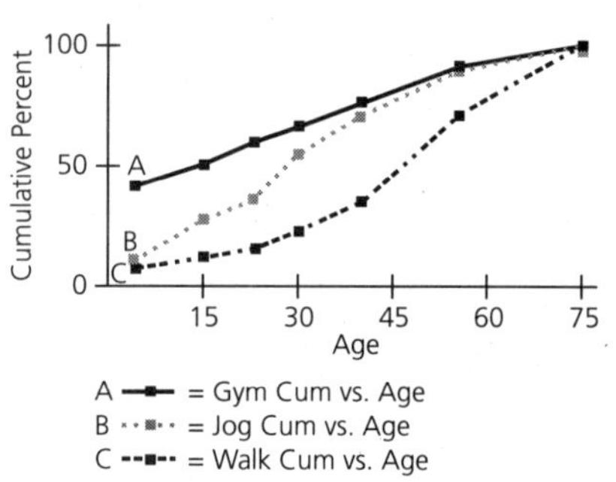

b. From the table or the cumulative-percent plot, about half of the gym-shoe buyers are under the age of 17; about half of the jogging-shoe buyers are under the age of 33—something a little less than 34; and about half of the walking-shoe buyers are under the age of 52—not quite half way between 44 and 64.

c and d. Using midpoints of the intervals as the value for the age of each group, the means and the standard deviations are approximately those in the following table.

	Mean	Standard deviation
Gym shoes	24.86	19.89
Jogging shoes	34.79	17.23
Walking shoes	51.23	18.55

Students may get quite different answers, depending upon how they choose the midpoint of the age groups and how they cut off the oldest age group. The mean is higher than the median for gym shoes, which has the age distribution with severe skewness toward the older age groups. The mean and median are close for the other two types of shoes, but remember that the calculations here are rough approximations to the actual values.

e. Of three types of shoes—gym, jogging, and walking—that make up a large part of the sport-shoe market, gym shoes have the

STUDENT PAGE 26

c. It is difficult to calculate the mean age of each user since the ages are given in intervals. For each age group, select a single value which you think best approximates the ages in that interval. Using those selected values, approximate the mean age of user for each of the three shoe types. How do the mean ages compare with the median ages?

d. Using the ages per interval selected in Problem 13c, approximate the standard deviation of ages for each of the three shoe types.

e. Would manufacturers of sports shoes find these means and standard deviations to be useful summaries of the age distributions? Write a summary of these age distributions for a publication on shoe sales, assuming the audience knows very little about statistics.

14. This lesson began with a discussion of the distribution of the number of pets found in a sample of students. In Lesson 3, we assumed that it cost $20 per week to feed each pet. The distribution of *Y*, the weekly cost of feeding pets, is shown in this table.

Y	*P*(*Y*)
0	0.2
20	0.5
40	0.3

a. Use this distribution to find the standard deviation of *Y*. You may use the fact that the expected value of *Y* is $22.

b. Compare the standard deviation of *Y* to the standard deviation of the number of pets per student, found in Problem 2 to be 0.70. Do you see a simple rule for relating the standard deviation of *Y* to that of *X*, the number of pets per student?

15. Suppose a random variable, *X*, has standard deviation denoted by σ, the Greek letter s. A new random variable is constructed as $Y = aX + b$.

a. What is the standard deviation of *Y* in terms of σ? Show why this is true by making use of the formula for standard deviation.

b. Suppose the number of persons per household has a mean of 2.6 and a standard deviation of 1.4. Each mem-

26 LESSON 4

youngest buyers and walking shoes the oldest. The average age of a gym-shoe buyer is approximately 25, but half of these buyers are 17 or younger. The typical purchaser of jogging shoes is about 35 years old, and these buyers are concentrated in an age range from 17 to 52. For walking shoes, the typical buyer is about 51 years old, with these buyers concentrated in an age range from 32 to 70.

14. a. $SD(Y) = 14$

b. $Y = 20X$; $SD(Y) = 20 \cdot SD(X) = 20(0.70) = 14$

15. a. This demonstration involves symbol manipulation, which is a big jump from working with data. Students may need some guidance.

$$V(Y) = \sum_{1=1}^{n} (y - \mu_y)^2 p_i$$

$$= \sum_{1=1}^{n} [(ax_i + b) - (a\mu_x + b)]^2 \, p_i$$

$$= \sum_{1=1}^{n} [a(x_i - \mu_x)]^2 \, p_i$$

$$= a^2 \sum_{1=1}^{n} [(x_i - \mu_x)]^2 \, p_i = a^2 \, [V(X)],$$

where μ_x denotes $E(X)$ and μ_y denotes $E(Y)$.

From this result, we see that $SD(Y) = \sqrt{V(Y)} = |a| \, [SD(X)]$.

b. With the number of persons in a randomly selected household denoted by Y and the cost of interviewing by C, $C = 30Y$, $E(C) = 30E(Y) = 30(2.6) = \$78$, and $SD(C) = 30SD(Y) = 30(1.4) = \$42$. Since \$100 is only about one half a standard deviation above the expected cost, it could be exceeded fairly often.

STUDENT PAGE 27

ber of a sampled household is to be interviewed by a pollster at a cost of \$30 per interview. What are the expected value and standard deviation of the cost of interviewing a randomly selected household? Would this cost exceed \$100 very often?

SUMMARY

For distributions of data and probability distributions of random variables the center is often measured by the mean or expected value and the spread by the ***standard deviation***. The standard deviation measures a "typical" deviation between a possible data point and the mean. Most of the data points usually lie within one standard deviation of the mean.

For probability distributions, the standard deviation can be written as an expected value of a function of the underlying random variable. This measure will be used extensively in future lessons of this module.

ASSESSMENT

Lessons 1–4

STUDENT PAGE 28

Solution Key

1. Possible answer: The distribution of number of persons per household has its greatest frequency at 2 persons per household, with 90% of the values concentrated from 1 through 4. The distribution is skewed toward larger families, with very few households having more than 7 persons. The distribution of number of children per family has its greatest frequency at 0 and 90% of the values are concentrated from 0 through 2. The distribution is skewed toward more children, with very few families having more than 4 children.

2. Possible answer: Suppose a random sample of households is to be selected from all U.S. households by a polling organization. Define the random variable T as the total number of persons per household among those households sampled. T will vary from household to household in the sample, but we anticipate that the distribution of values for T should look much like the table in Problem 1 after the sample is selected and measured.

3. **a.** $1 - 0.25 = 0.75$
   b. $1 - 0.25 - 0.32 = 0.43$
   c. $1 - 0.25 - 0.32 = 0.43$
   d. $0.25 + 0.32 + 0.17 = 0.74$
   e. $0.32 + 0.17 + 0.16 = 0.65$
   f. $0.32 + 0.17 + 0.16 + 0.07 = 0.72 = 1 - 0.28$

4. **a.** $P(T \geq 2) = 0.75$
   b. $P(T > 2) = 0.43$
   c. $P(T \geq 3) = 0.43$
   d. $P(T \leq 3) = 0.74$
   e. $P(2 \leq T \leq 4) = 0.65$
   f. $P(T > 1 \text{ and } T < 6) = P(1 < T < 6) = P(2 \leq T \leq 5) = 0.72$

ASSESSMENT

Lessons 1–4

1. The following table shows the distribution of family sizes for U.S. families.

Number of Persons per Household	Percent of Households
1	25
2	32
3	17
4	16
5	7
6	2
7	1

Note: Households of more than 7 persons are very rare.

Describe the distribution of household size and compare it to the distribution of number of children per family.

2. Provide a reasonable definition for a random variable whose distribution can be approximated from these data on number of persons per household.

3. For a randomly selected family from the U.S., find the probability that the number of persons in the household is
   a. 2 or more.
   b. more than 2.
   c. at least 3.
   d. no more than 3.
   e. between 2 and 4, inclusive.
   f. more than 1, but less than 6.

4. Write each of the statements in Problem 3 in symbolic form.

5. According to the U.S. Bureau of the Census, the number of cars available to American households is given by the following percents.

28 ASSESSMENT

5. a. $E(C) = 2.17$

b. $SD(C) = 0.94$

c. $1000(2.17) = 2170$

d. $\frac{1000}{2.17} \approx 461$

e. $\$250(2.17) \approx \$542.50$

f. $\$250(0.94) = \$235.00$

g. E(Cost for 1000 households) $= 1000(542.50) = \$542{,}500$; SD(Cost for 1000 households) $= 1000(235) = \$235{,}000$

No; the actual cost for maintenance by 1000 families could easily deviate from the expected cost by a standard deviation or more.

STUDENT PAGE 29

Number of Cars per Household	Percent of Households
0	1.4
1	22.8
2	43.7
3	21.5
4	10.6

A random sample of households is to be selected to participate in a study entitled "How much do you spend on auto repairs?"

a. What is the expected number of cars per randomly selected household?

b. What is the standard deviation of the probability distribution for the number of cars per randomly sampled household?

c. If the poll selects 1000 households, how many cars are expected to be represented in the poll?

d. If the polling organization expects 1000 cars to be represented, how many households should be sampled?

e. Car maintenance costs average $250 per year, not counting gas and oil. How much would a randomly selected household expect to pay annually for car maintenance?

f. What is the standard deviation of the amount a randomly sampled household is to pay for car maintenance?

g. Use the information in Problem 5e to determine the total amount a random sample of 1000 families would expect to pay for car maintenance in a year. Do you think this expected value will be a good approximation to the real total for the 1000 families? (HINT: Find the standard deviation of the total amount 1000 families might have to pay for car maintenance and use that value in your answer.)

6. The following table shows the age distribution of residents of the United States for the years 1990 and 1996, according to the U.S. Bureau of the Census. The population figures are given in thousands. The columns labeled "proportion" show the proportions of the residents in each of the age categories.

6. a.

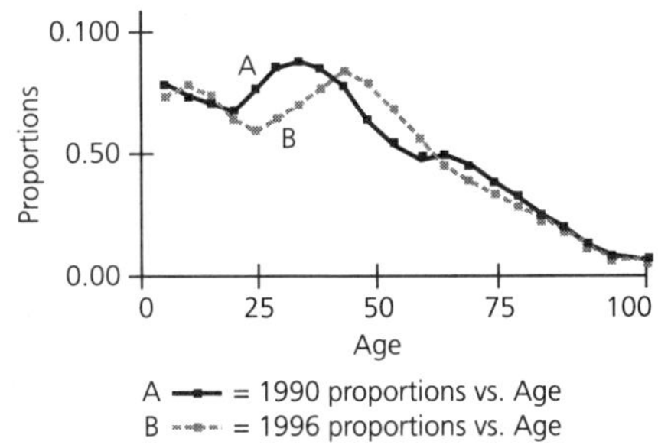

The graph shows the proportions as a function of the midpoint of each age group, with 102.5 used as the midpoint of the oldest group.

b. The key difference between the two distributions is the shift that occurs from the teens through the 40s. The 1990 population has a large group in the teens, 20s, and 30s that by 1996 has moved over to the 20s, 30s, and 40s. Even in the short time of six years, significant shifts in the age distribution of the population can be seen.

c. The median ages are approximately 32 and 34 for 1990 and 1996, respectively. As should be the case, 1996 has the greater median.

d. The expected age for 1996 is approximately 35.96. The expected age for 1990 is approximately 35.26. Both of the means are greater than the respective medians, which is not surprising because of the skewness of the age distributions toward the older ages.

The mean ages for the two years are very close, since the extremes of the two distributions are very nearly identical. The medians differ a little more because of the shifting near the centers of the distributions.

e. The standard deviation for the 1996 data is approximately 22.4. For the 1990 data, it is approximately 22.2, again almost identical. This is a reasonably good approximation of "typical" deviations from the mean. Most of the ages are, in fact, between 36 − 22 = 14 and 36 + 22 = 58.

STUDENT PAGE 30

	Age	1990	1990 Proportion	1996	1996 Proportion
1	Under 5	18,849	0.076	19,354	0.073
2	5 to 9	18,062	0.072	19,640	0.074
3	10 to 14	17,189	0.069	19,131	0.072
4	15 to 19	17,750	0.071	18,699	0.070
5	20 to 24	19,135	0.077	17,307	0.065
6	25 to 29	21,233	0.085	19,004	0.071
7	30 to 34	21,906	0.088	21,217	0.080
8	35 to 39	19,975	0.080	22,508	0.085
9	40 to 44	17,790	0.071	20,940	0.079
10	45 to 49	13,820	0.055	18,474	0.069
11	50 to 54	11,368	0.046	14,216	0.053
12	55 to 59	10,473	0.042	11,429	0.043
13	60 to 64	10,619	0.042	9,997	0.038
14	65 to 69	10,077	0.040	9,873	0.037
15	70 to 74	8,022	0.032	8,773	0.033
16	75 to 79	6,145	0.025	6,928	0.026
17	80 to 84	3,934	0.016	4,587	0.017
18	85 to 89	2,049	0.008	2,399	0.009
19	90 to 94	764	0.003	1,020	0.004
20	95 to 99	207	0.001	288	0.001
21	100 or more	37	0.000	58	0.000

a. Show appropriate plots of the age distribution for 1990 and the age distribution for 1996.

b. Discuss the key differences between the shapes of the two age distributions. What is the major change in the age distribution between 1990 and 1996?

c. Approximate the median age for the 1990 population. Do the same for the 1996 population. Compare the median ages.

d. Suppose the Gallup organization is to take a random sample of a large number of residents of the U.S. What can they expect as the mean age of those in their sample? How does this expected value compare to the median found above? How does this expected value compare to a similar expectation found for a sample taken in 1990?

e. Under the conditions described in Problem 6d, what can the Gallup organization expect as the standard deviation of the ages of the people who end up in their random sample? In this case, is the standard deviation a good description of a "typical" deviation from the mean age?

30 ASSESSMENT

STUDENT PAGE 31

7. Possible answer: From the percents given in the data, we can arrive at the following percents for the four classes of income data.

Education	Percent	Typical income
Less than high school	10	14,000
High-school degree	33	20,000
Some college	28	23,000
College degree	29	43,000

From this table, E(Income) = \$26,910. This expected value is based on very broad categories of workers and rough approximations to typical incomes for each category. More detailed data could give quite a different expected value. For example, professionals with advanced degrees have much greater typical incomes than \$43,000 per year. The expected value could increase if these incomes were a direct part of the calculation.

7. According to Census data, about 90% of the U.S. work force have at least a high-school education, about 57% have at least some college education, and about 29% have at least a bachelor's degree from a college or university. Suppose a typical worker without a high-school education earns about \$14,000 per year, a typical high-school graduate makes about \$20,000 a year, a typical worker with some college experience but not a bachelor's degree makes about \$23,000 a year, and a typical worker with at least a bachelor's degree makes about \$43,000 per year. Find the expected yearly income for a person randomly selected from the U.S. workforce. Explain why this expected value may be slightly different from the true mean income of the U.S. workforce.

UNIT II

Sampling Distributions of Means and Proportions

LESSON 5

The Distribution of a Sample Mean

Materials: 100 chips or small equal-sized pieces of paper for each group of students. Let the students put the numbers on the chips so that they can clearly understand every step of the simulation.
Technology: graphing calculators or computers (optional)
Pacing: 1 class period or more

Overview

A fundamental principle of statistics is that in random sampling the potential values of a sample mean have approximately a mound-shaped symmetric distribution centering at the population mean and with standard deviation equal to the population standard deviation divided by the square root of the sample size. This result, known as the "Central Limit Theorem," is best discovered by students through simulation. The sampling distribution of the mean can be called "normal," but details on the normal distribution are deferred until the next lesson.

Teaching Notes

Allow enough time for small groups of students to work through the simulation by hand, so that they can see the patterns in the distributions of the sample means unfold. You might have students do the simulation and collect the data in one session and then report the results at the beginning of the next; or the students could conduct the simulation outside of class. Organize the class discussion so that the sample results from the groups can be quickly collected and assembled for display. The dotplots (number-line plots) or stemplots of the collected means can be made on an overhead transparency or with a computer or graphing calculator with display capabilities.

It is important for students to see that the shape of the simulated sampling distribution gets more symmetric and mound-shaped as the sample size increases. Also, the mean of the sampling distribution does not change, but the variation decreases as the sample size increases. Point out that the theoretical formula for the standard deviation of the sampling distribution appears to work quite well. This result allows us to anticipate the distribution of possible values of a sample mean in random sampling quite precisely and is one of the most amazing results in probability and statistics.

Technology

The simulation here should be done by hand. A graphing calculator or computer may be used to help display the data and to calculate the summary statistics for the collected means.

Follow-Up

Another simulation of the sampling distribution for the sample mean can be done, using another of the distributions from earlier lessons, such as the number of cars per household, or a distribution of the students' choice. The second time through, students could generate the simulated sample means using a computer or graphing calculator.

STUDENT PAGE 35

LESSON 5

The Distribution of a Sample Mean

If you were to sample 100 families, what is the total number of children you would expect to see?

What is the distribution of potential values of the mean number of children per family in the sample of 100 families?

How will the distribution of potential values of the mean change with the sample size?

What is the average summer daytime temperature for your town? What is the average age of a student in your class? What is the average number of points scored by your basketball team during the season? What is the average time it takes you to get to school in the morning? Averages used as summary or typical numerical values are all around us. When working with data, we have seen that the arithmetic average is called the mean. When working with a probability distribution, a possible model for data yet to come, the average is called the expected value.

OBJECTIVE

Understand the behavior of the distribution of means from random samples.

Much of the data we see in designed studies, such as sample surveys, comes about through random samples from specific populations. Since these data are typically reported in summary form as means, it is important that we understand the behavior of sample means that arise from random sampling.

INVESTIGATE

The A. C. Nielsen Company samples households to collect data on TV-viewing habits. For some shows, the company is particularly interested in the number of children under the age of 18

Solution Key

Discussion and Practice

1. We expect to see only 910 children in a sample of 1000, so the chance of seeing more than 1000 does not appear to be too great. Answers will vary, but students generally don't think things will deviate far from the mean or expected value.

2. Now, we expect to see 1092 children in a sample of 1200, so the chance of seeing more than 1000 appears to be quite good. In fact, it seems like a virtual certainty.

STUDENT PAGE 36

who might be watching. Thus, it wants to make sure that there will be a reasonable number of children in its sample of households. One way to predict where this number might lie is to study the possible values of the mean number of children per sampled family in a typical random sample of families.

Something is known about the number of children in U.S. families, and that is the place to begin. The available population information comes in the form of the Census Bureau's distribution of children per family, as used in earlier lessons. The data are in the table below.

Number of Children	Percent of Families
0	51
1	20
2	19
3	7
4	3

An approximation to the expected number of children per family, as calculated in earlier lessons, is approximately 0.91.

From information gathered so far, Nielsen can tell something about the expected number of children per family in a random sample of n families. What other information does the company need?

Discussion and Practice

Nielsen wants to choose a sample size that produces a reasonable number of children with high probability.

1. If $n = 1000$, how many children would Nielsen expect to see in the sample of families? Suppose the company has a goal of seeing at least 1000 children in its survey. Do you think the probability of seeing more than 1000 children is high for a sample of 1000 families?

2. Suppose the sample size is increased to $n = 1200$. Will that increase the probability that Nielsen will achieve the goal of at least 1000 children in the sample? Will that probability change a great deal or very little?

In order to get specific answers to questions like those just posed, more information on the probability distribution of sample means from random samples must be developed; that is

36 LESSON 5

3.–4. An example of the results from a simulation like the one described here is provided in Figure 1 of Problem 6 in the student text. The sampling distributions for the means are mound-shaped and nearly symmetric. They center at about 0.91, the expected value for the original distribution of children per family, and have variation that decreases as the sample size increases.

STUDENT PAGE 37

the goal of this lesson. This new information will be discovered through simulation.

The Sampling Distribution of a Mean

If one family is randomly selected from the U.S. population, what is the probability that the family will contain no children? One child? Four children? Our first job is to design a simulation for random sampling that will preserve these probabilities, yet allow us to look at typical sample data on children per family. This investigation can be completed most efficiently if you work in small groups and then combine data for the class.

3. Designing and conducting the simulation

a. Each group must have 100 plastic chips or small pieces of paper of equal size. Number the chips with integers 0 through 4, in the same percents given in the table for number of children per family, and place the chips into a box. You have now constructed a physical model of the probability distribution of the number of children per family. If you reach into the box of chips and randomly select one chip, what is the probability that it will show a 2? Is this the probability you want for modeling Nielsen's sampling process?

b. Select a simulated sample of 10 households. Randomly draw a series of 10 chips from the box, but be careful to *replace* each sampled chip before the next one is drawn. Why is it important to do the sampling with replacement?

c. What percent of the sample outcomes were *zeros*? What percent were *ones*? Is this approximately what you expected?

d. Calculate the mean number of children per household found in your random sample of 10 households.

e. Repeat the process for three more samples of 10 households each. Record and save the sample data and the values of the means. Your group should now have four samples.

f. Collect the sample means produced by all the groups in the class. Plot the collection of sample means on a dotplot or stemplot so that the shape of the distribution can be seen. Comment on this shape. Does it differ from the

STUDENT PAGE 38

5. The mean of each simulated sampling distribution should be close to 0.91. The standard deviation of each sampling distribution should be close to 1.114, the standard deviation of the original distribution, divided by the square root of the sample size. This result is demonstrated in Problem 6 in the student text.

You might ask students to investigate the relationship between the standard deviation of the sampling distribution and the sample size graphically. By plotting observed *SD* against sample size as in the first plot below, we can see that we have a type of inverse relationship. With a little investigation, students can discover that a straight line is produced by plotting observed *SD* against the inverse of the square root of the sample size as in the second plot below. The straight line here has a slope a little greater than 1, about the size of the population standard deviation.

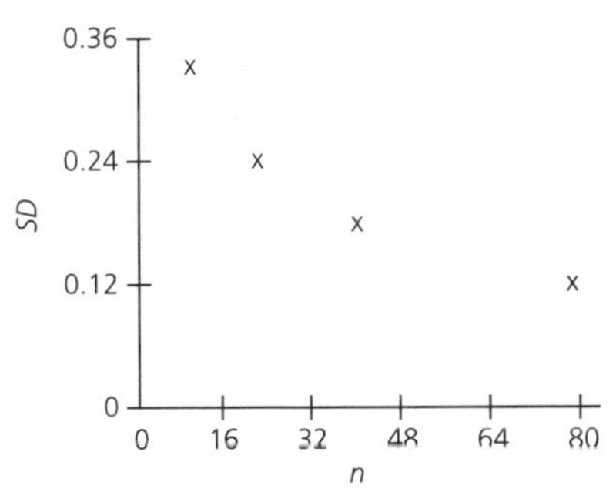

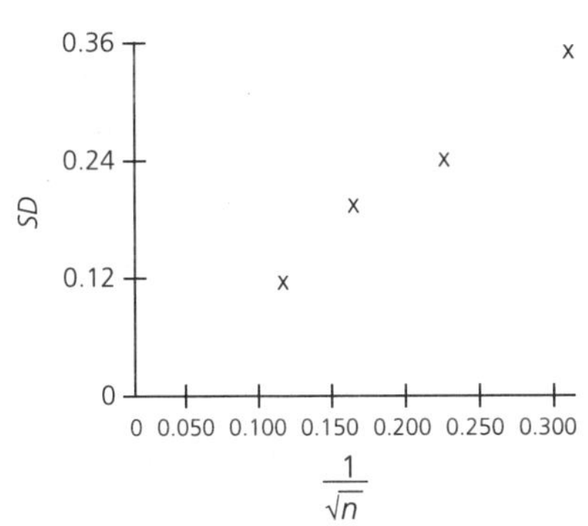

shape of the population distribution of children per family? How does it differ?

4. Changing the sample size

a. Repeat the simulation of Problem 3 for samples of 20 households each. This can be accomplished by making two pairs of the size-10 samples already selected, yielding two size-20 samples. Collect the sample means produced by all groups in the class, plot the means on a dotplot or stem-and-leaf plot, and comment on the shape.

b. Repeat the simulation of Problem 3 for samples of 40 households each. This can be accomplished by combining the two size-20 samples already available in each group. Collect the sample means produced by all groups in the class, plot the means on a dotplot or stem-and-leaf plot, and comment on the shape.

c. Repeat the simulation of Problem 3 for samples of 80 households each. This can be accomplished by combining the 40 data points from your group with 40 from another group. Make sure each group's data is used only once. Collect the sample means produced by the class, plot the means on a dotplot or stem-and-leaf plot, and comment on the shape.

d. The distribution of possible values of the sample mean is called a ***sampling distribution***. Study the plots of sampling distributions for size-10, size-20, size-40, and size-80 samples, and comment on

i. the shapes of the sampling distributions of sample means.

ii. the centers of the sampling distributions of sample means.

iii. the variation in the sampling distributions of sample means.

5. Computing summary statistics

a. Calculate the means and the standard deviations for each of the simulated sampling distributions produced above. That is, use the original sets of sample means generated by the class in the simulations to calculate the mean and the standard deviation of the sample means within each sample size.

38 LESSON 5

STUDENT PAGE 39

b. How do the calculated means of the simulated sampling distributions compare to the expected number of children per family (the mean of the population) calculated from the population distribution to be $\mu = 0.91$? (The symbol μ used for the population mean is the Greek mu, or m.) Make a general statement about how the means of the sampling distributions relate to the mean of the population from which the samples were selected.

c. How do the calculated standard deviations of the simulated sampling distributions compare to the standard deviation of the population, calculated to be $\sigma = 1.114$? (The symbol σ used for the population standard deviation is the Greek sigma, or s.) Do you see a pattern developing in how the standard deviations of the sampling distributions relate to the sample sizes?

d. The precise relationship between the standard deviation of a sampling distribution for means and the sample size is difficult to see intuitively, so we'll provide some help. We denote the population standard deviation by σ and label the standard deviation of a sampling distribution of sample means by *SD*(mean). You have noticed that *SD*(mean) decreases as the sample size *n* increases. Mathematical theory of statistics says that the precise relationship among these quantities is given by

$$SD(\text{mean}) = \frac{\sigma}{\sqrt{n}}$$

For $\sigma = 1.114$, as it is for the population of number of children per household, calculate the theoretical *SD*(mean) for sample sizes of 10, 20, 40, and 80. Compare these theoretical values to the observed standard deviations of the sampling distributions calculated in Problem 5a.

PATTERNS

To show that the patterns generated in sampling distributions simulated by your class are not accidental, and to show a slightly larger simulation, we have produced a computerized version of the simulation outlined above. In this case, we have calculated the means from 100 random samples in each simulated sampling distribution. The results are shown in the dotplots that follow.

STUDENT PAGE 40

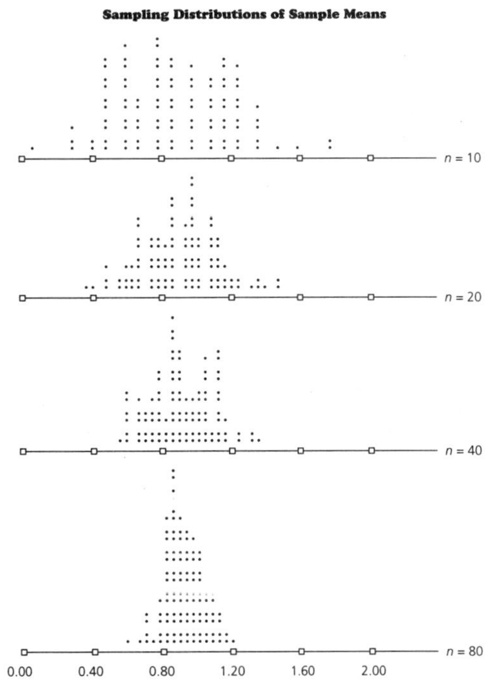

	No. Samples	Mean	Standard Deviation
$n = 10$	100	0.9140	0.3426
$n = 20$	100	0.9320	0.2397
$n = 40$	100	0.8970	0.1891
$n = 80$	100	0.9165	0.1188

Notice three things:

- The sampling distributions all center around the population mean of 0.91.
- The standard deviations of the sampling distributions get smaller as the sample size increases.

40 LESSON 5

Practice and Applications

6. **a.** For a sample of 10 families to contain at least 30 children, the sample mean would have to exceed 3. From the first graph in the Summary following Problem 5, we see that this never happens.

b. For a sample of 20 families to contain at least 30 children, the sample mean would have to exceed 1.5. From the second graph in the Summary, we see that this could happen but has only a small chance.

c. For a sample of 40 families to contain at least 30 children, the sample mean would have to exceed 0.75. From the third graph in the Summary, we see that this happens with probability greater than 0.5.

d. For a sample of 80 families to contain at least 30 children, the sample mean would have to exceed 0.375. From the fourth graph in the Summary, we see that happens virtually all the time.

7. **a.** In a random sample of 1000 families, the potential values of the sample mean would show a mound-shaped, symmetric distribution centering at 0.91, the expected number of children per family, and with a standard deviation of 0.035. The latter implies that more than half of the time the sample means would lie between 0.875 and 0.945. So, the distribution of potential values of the sample mean is concentrated very tightly around the expected value.

STUDENT PAGE 41

- The sampling distributions tend to have somewhat symmetric mound shapes.

To be more specific about the second point, let's compare the observed standard deviations of the sampling distributions to the theoretical values that should be generated:

n	*SD* observed	*SD*(mean) = $\frac{\sigma}{\sqrt{n}}$
10	0.3426	0.3523
20	0.2397	0.2491
40	0.1891	0.1761
80	0.1188	0.12457

It appears that the theoretical rule for relating the standard deviation of the sampling distribution to the population standard deviation and the sample size works well. We will return to the point about the symmetric, mound-shaped distribution in the next lesson.

Practice and Applications

6. Suppose the Nielsen Company wants a sample of families containing at least 30 children in all. Is this highly likely with a random sample of the size given?

a. $n = 10$

b. $n = 20$

c. $n = 40$

d. $n = 80$

Explain how to use the simulated sampling distributions on page 40 to answer this question.

7. Suppose the Nielsen company is to select a random sample of 1000 families.

a. Describe the distribution of potential values of the sample mean number of children per family. The description should include a statement about the center and spread of the distribution of potential values.

b. If Nielsen wants to see at least 1000 children in the sample, what would the mean number of children per family have to equal or exceed? Do you think it is likely that the sample of 1000 households will produce at least 1000 children? (HINT: It is unusual for a data value to

THE DISTRIBUTION OF A SAMPLE MEAN **41**

b. With the distribution of sample means showing an expected value of 0.91 and a standard deviation of 0.035, it follows that a mean of 1 child per sampled family is 2.57 standard deviations above the mean of the distribution. A value greater than a point this far above the center of the distribution has a very small chance of occurring.

c. With the distribution of sample means showing an expected value of 0.91 and a standard deviation of 0.035, it follows that a mean of $\frac{1000}{1200} = 0.83$ child per sampled family is more than 2 standard deviations *below* the mean of the distribution. A value greater than a point this far below the center of the distribution has a very great chance of occurring; in fact, such a value will occur nearly every time.

8. We need to find a mean weight for the random sample of n people who enter the elevator on any given run and choose n to make sure this mean does not exceed $\frac{2000}{n}$ very often.

Suppose $n = 14$. Then $\frac{2000}{n} =$ 142.8, and the standard deviation of the distribution of sample means is $\frac{10}{\sqrt{n}} = 2.67$. Since this value for the sample mean is below the expected value of the distribution of sample means (given as 150) it will be exceeded with probability greater than 0.5. This n is too large for safe running of the elevator.

Suppose $n = 13$. Then $\frac{2000}{n} =$ 153.8, and the standard deviation of the distribution of sample means is $\frac{10}{\sqrt{n}} = 2.77$. Since this value for the sample mean is about 1.37 standard deviations above the expected value of the distribution of sample means (given as 150) it will be exceeded with probability much less than 0.5. The weight limit will be exceeded on occasion, but this may allow for safe running of the elevator.

Suppose $n = 12$. Then $\frac{2000}{n} =$ 166.7, and the standard deviation

STUDENT PAGE 42

lie more than two standard deviations from the mean of the distribution from which it was selected.)

c. Suppose Nielsen changes to a random sample of 1200 households. Does this dramatically improve the chance of seeing at least 1000 children in the sampled households? Explain.

8. The people using an elevator in an office building have an average weight of approximately 150 pounds and a standard deviation of weights of approximately 10 pounds. The elevator is designed for a 2000-pound weight maximum. This maximum can be exceeded on occasion, but should not be exceeded on a regular basis. Your job is to post a sign in the elevator stating the maximum number of people for safe use. Keep in mind that it is inefficient to make this number too small, but dangerous to make it too large. What number would you use for maximum occupancy? Explain your reasoning.

9. A call-in radio show collects callers' opinions on the number of days students should be in school during a year. The mean number for 500 callers was 195 days. The radio show then announces that this mean should be close to the mean one would obtain if all residents of the community were asked this question. What is wrong with this reasoning?

10. What happens to the mean of the sampling distributions as the sample size increases, everything else remaining fixed? How does the mean of the sampling distributions compare to the mean of the population from which the samples were selected?

11. What happens to the standard deviation of the sampling distributions as the sample size increases, everything else remaining fixed? How does the standard deviation of each of the sampling distributions compare with the standard deviation of the population from which the samples were selected?

SUMMARY

Means, or averages, are one of the most common summary statistics used to describe data. Thus, to make inferences from data we must understand how means of random samples behave. The distribution of potential values of the sample mean to be produced by a random sample from a fixed population is called a ***sampling distribution***. Sampling distributions for

42 LESSON 5

of the distribution of sample means is $\frac{10}{\sqrt{n}} = 2.89$. Since this value for the sample mean is about 5.6 standard deviations above the expected value of the distribution of sample means (given as 150) it will be exceeded almost never. This would be a safe way to run the elevator, but it may be too conservative, as the elevator may then be forced to make more trips than necessary.

9. Possible answer: Those who choose to call a radio show are not a random sample of the population at large. Thus, there is no way to tell how far this mean may actually be from the true population mean. The methods of this lesson will not work with this type of non-random data.

10. The mean of the sampling distribution of sample means should always be equal to the mean of the population from which the samples are selected, regardless of the sample size.

11. The standard deviation of the sampling distribution of sample means is equal to the population standard deviation divided by the square root of the sample size. This implies that the spread of the sampling distribution decreases as the sample size increases.

STUDENT PAGE 43

means have three important properties:

- The sampling distributions all center around the population mean.
- The standard deviations of the sampling distributions get smaller as the sample size increases, and this can be predicted by the rule

$$SD(\text{mean}) = \frac{\sigma}{\sqrt{n}}$$

- The sampling distributions tend to be symmetric and mound-shaped.

The first two properties were investigated in this lesson; the third is the subject of the next lesson. This result is commonly known as the "Central Limit Theorem."

THE DISTRIBUTION OF A SAMPLE MEAN **43**

LESSON 6

The Normal Distribution

Materials: random-number generator
Technology: graphing calculators or computers (optional)
Pacing: 1 class period

Overview

The normal distribution is the most common probability model used in statistics because it is the approximate sampling distribution of sample means and sample proportions. This lesson deals with its basic properties, including the fact there is a relationship between intervals on both sides of the mean and the relative frequency with which values will fall into those intervals. The mean is a very useful summary statistic, and the exercises are designed to show how knowledge of the mean's sampling distribution can lead to intelligent decisions in a variety of contexts.

Teaching Notes

Another simulation is introduced here, this time with random digits, to ensure that students grasp the idea that the normality of the sampling distribution of sample means is a general result. Students should see that the relative frequencies associated with the normal distribution allow us to state that the sample mean will be within two standard deviations of the population mean about 95% of the time. This is a very useful result for making practical decisions, as in the elevator problem.

Technology

Since the simulation of this lesson is based on random digits, it can be done with a graphing calculator or computer.

Follow-Up

The lesson deals with only the basic properties of the normal distribution and looks at probabilities in intervals of one or two standard deviations to either side of the mean. Since most graphing calculators have the normal distribution built in, this can be generalized to consider problems of any fractional distance to either side of the mean.

STUDENT PAGE 44

LESSON 6

The Normal Distribution

What is the chance that your school's mean weekly earnings from recycling aluminum cans during the fall will exceed $130?

Do the probability distributions of potential values of a sample mean always have nearly the same shape?

How can you make use of a common model for distributions of means?

OBJECTIVES

Understand the basic properties of the normal distribution.

See the usefulness of the normal distribution as a model for sampling distributions.

Means are widely used statistical summaries, and decisions based on means can be more enlightened if decision-makers understand the behavior of sample means from random samples. Suppose records show that the weekly amounts your school earns on recycling aluminum cans has a mean of $120 and a standard deviation of $8. During a sixteen-week period in the fall, what is the chance that the mean weekly earnings will exceed $130? $124? $100? In Lesson 5, you used simulation to answer questions similar to these. But it is cumbersome and time-consuming to conduct a simulation every time you want to answer a question about a potential value of a mean. It would be helpful to have a ***model*** for the behavior of sample means which would give quick approximate answers to the many questions that arise about sample means. Such a model is the ***normal distribution.***

For the A. C. Nielsen Company, it may be important to know even more details than provided in Lesson 5 about the possible values of mean number of children per sampled family. One such question might be, "What is the chance of having fewer than 1000 children in a sample of 1200 families?" Relating the

44 LESSON 6

STUDENT PAGE 45

sampling distributions discovered in Lesson 5 to a theoretical model for such distributions—the normal distribution—can help provide answers to such questions.

INVESTIGATE

Consider, once again, the plots constructed in Lesson 5 (on page 40) that show the probabilistic behavior of sample means for various sample sizes. These curves look fairly symmetric and mound-shaped. Such mound-shaped, symmetric distributions are seen very often in the practice of plotting data. In fact, they are seen so often that such a curve is called "normal," and a theoretical model for this curve has been studied extensively.

Normal curves have two key measurements that determine their location and shape. One, the location of the center of the curve on the real-number line, is the mean, usually denoted by m. The other, the measure of spread, or width, of the curve, is the standard deviation, usually denoted by s. Pictures of two normal curves with different standard deviations are shown below. Note that standard deviation measures variability. A careful look at these pictures shows the standard deviation to be half of the curve about $\frac{2}{3}$ of the way up the line of symmetry, the vertical line through the mean.

Normal Distributions

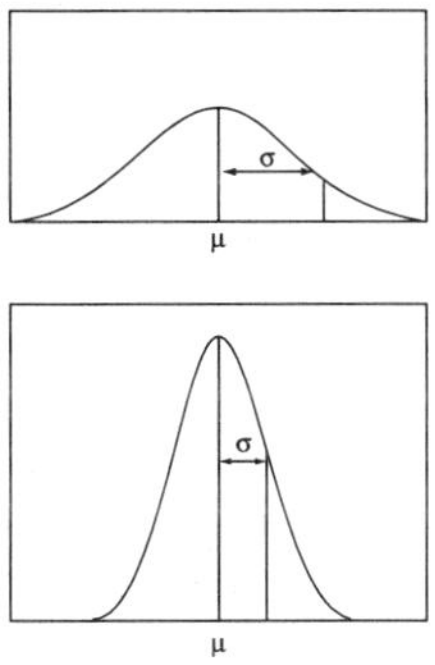

The area under the curve over an interval on the horizontal axis represents the percent of the data that fall into that interval. These intervals can be located in terms of the mean and

Solution Key

Discussion and Practice

1. **a.** 50%

b. $\frac{100-68}{2} = 16\%$

c. $34 + 47.5 = 81.5\%$

d. $100 - 95 = 5\%$

2. Possible answer: No; there is a high percent of families with no children, and then the percents decrease as the number of children increases. The distribution is strongly skewed toward more children.

STUDENT PAGE 46

standard deviation of the distribution of data. For the normal curve, it is common practice to describe the distribution in terms of intervals that are symmetric about the mean. For the normal distribution,

the interval $\mu \pm \sigma$ contains about 68% of the data,
the interval $\mu \pm 2\sigma$ contains about 95% of the data,
and the interval $\mu \pm 3\sigma$ contains about 99.7% of the data.

These intervals and their respective areas are shown in the graph that follows. The scale on the horizontal axis is in terms of standard-deviation units. A point at 1 is one standard deviation above the mean. A point at –2 is two standard deviations below the mean.

Areas Under a Normal Distribution

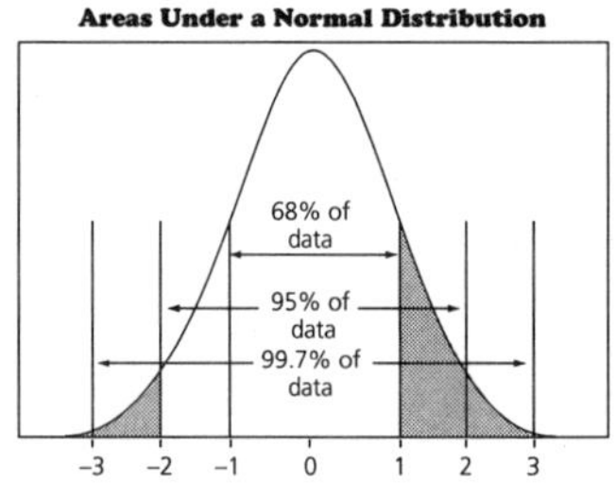

Discussion and Practice

1. For a distribution of data that can be represented by a normal curve,

a. what percent of the data is below the mean?

b. what percent of the data is more than one standard deviation above the mean?

c. what percent of the data is between one standard deviation below the mean and two standard deviations above the mean?

d. what percent of the data is more than two standard deviations away from the mean?

2. Does the population distribution of number of children per family (See Lesson 1 or the Assessment for Lessons 1–4.) look "normal"? Explain your reasoning.

46 LESSON 6

STUDENT PAGE 47

3. Recall other distributions of data you have seen in these lessons or other places. Describe at least two other data sets that look normal to you. Explain your reasoning.

The Normal Distribution as a Model

The task at hand is to relate the normal-looking sampling distributions for the sample mean, found in Lesson 5, to the normal distribution. The normal distribution, we discovered above, depends upon two constants, a mean and a standard deviation. Each of the sampling distributions seems to have approximately the same mean, and it is close to the population mean, or expected value, of 0.91 in the case of number of children per family. Therefore, we can take $\mu = 0.91$ in our theoretical normal model.

What about the standard deviation for the normal model? The standard deviations within the sampling distributions decrease as n increases, so each sample size generates a slightly different sampling distribution. These standard deviations are related to the population standard deviation by the equation

$$SD(\text{mean}) = \frac{\sigma}{\sqrt{n}}$$

where σ denotes the theoretical standard deviation of the underlying population. This relationship was explored in Lesson 5, with $\sigma = 1.114$ for the distribution of number of children per family.

4. Summarize the above information in a concise statement about the approximate sampling distribution of sample means for random samples of size n from a fixed population, by answering the questions below.
   a. What is the shape of the sampling distribution?
   b. Where is the center of the sampling distribution?
   c. What is the standard deviation of the sampling distribution?
5. Rewrite the basic relative-frequency rules for the normal distribution, preceding Problem 1 above, in terms of sampling distributions for sample means.
6. Look carefully at the distributions on page 40 of Lesson 5. What fraction of the observed sample means lie within two standard deviations of the mean of their distribution in each of the four cases? Do these fractions agree with what the normal distribution would predict?

THE NORMAL DISTRIBUTION 47

3. Possible answer: The distributions of number of cars per household in Lesson 1 and ages of purchasers of jogging shoes in Lesson 4 are somewhat mound-shaped and symmetric, similar to a normal distribution. Students may have seen distributions of measurements on weights or heights of men or women, which tend to be approximately normal in their distribution. Standardized exam scores, like those of the SAT, tend to be normally distributed. Many other examples are possible.

4. The sampling distribution for sample means, in random samples from a fixed population, have a distribution which is approximately normal in shape, centering at the true population mean and with standard deviation equal to the population standard deviation divided by the square root of the sample size. Recall that this result is known as the Central Limit Theorem.

5. The interval $\mu \pm \frac{\sigma}{\sqrt{n}}$ will contain about 68% of the possible values of the sample mean.

   The interval $\mu \pm 2\frac{\sigma}{\sqrt{n}}$ will contain about 95% of the possible values of the sample mean.

   The interval $\mu \pm 3\frac{\sigma}{\sqrt{n}}$ will contain about 99.7% of the possible values of the sample mean.

6. Shown on the following page are the stem-and-leaf or stemplots of the simulated sample means from Lesson 5. It is easier to count the exact number of values in prescribed intervals from this view of the data.

   The observed mean and the standard deviation are used in the calculation of the intervals here.

$n = 10$: 0.914 ± 2(0.3426) or (0.229, 1.599) contains 96% of the observed data values.

$n = 20$: 0.932 ± 2(0.2397) or (0.453, 1.411) contains 95% of the observed data values.

$n = 40$: 0.897 ± 2(0.1891) or (0.519, 1.275) contains 97% of the observed data values.

$n = 80$: 0.916 ± 2(0.1188) or (0.678, 1.154) contains 94% of the observed data values.

All of these percents agree closely with the 95% that the normal distribution would predict for these intervals.

Stemplot of $n = 10$ Leaf Unit = 0.010

1	1	0
1	2	
4	3	000
6	4	00
16	5	0000000000
27	6	00000000000
33	7	000000
45	8	000000000000
(10)	9	0000000000
45	10	000000000
36	11	00000000
28	12	0000000000
18	13	000000000
9	14	00000
4	15	0
3	16	0
2	17	
2	18	00

Stemplot of $n = 20$ Leaf Unit = 0.010

1	3	5
2	4	0
7	5	00055
13	6	000555
27	7	00000000555555
38	8	00000055555
(17)	9	00000000005555555
45	10	000000000000555555
27	11	00000000555555
13	12	00055
8	13	005
5	14	005
2	15	00

Stemplot of $n = 40$ Leaf Unit = 0.010

1	5	2
7	5	557777
12	6	22222
14	6	77
21	7	0022222
29	7	55557777
37	8	00022222
(15)	8	555555557777777
48	9	00022222
40	9	55577
35	10	00000022222
24	10	555577
18	11	0000222222
8	11	557
5	12	
5	12	55
3	13	22
1	13	5

Stemplot of $n = 80$ Leaf Unit = 0.010

1	5	6
1	6	
5	6	5678
7	7	01
12	7	56678
22	8	0000222233
44	8	5555555666666777788888
(16)	9	0001112222223333
40	9	5556667777788
27	10	0000011122233
14	10	5666777
7	11	0112
3	11	568

7. One possible simulation of sample means for samples of size 10 from random digits is shown below. Both the dotplot and stemplot are provided, along with the summary statistics.

DotPlot 'Means'

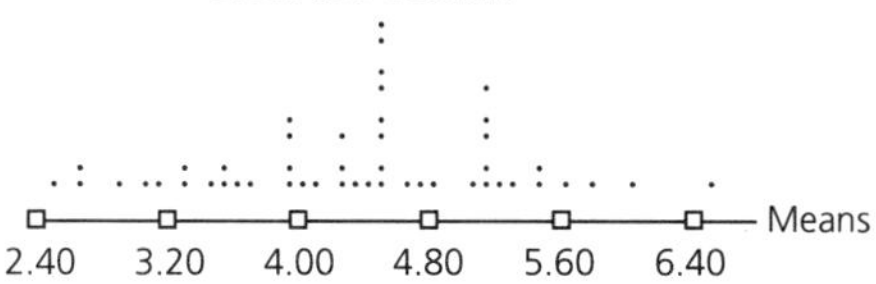

Stemplot of Means   Leaf Unit = 0.10

4	2	5669
9	3	01334
17	3	55679999
24	4	0122234
(11)	4	55555555678
15	5	0111112344
5	5	78
3	6	02
1	6	6

	n	Mean	Median	Standard Deviation
Means	50	4.364	4.500	0.952

The distribution is mound-shaped and somewhat symmetric, but we will not see a perfect normal curve for a sample size of 10 and only 50 means in the simulation.

For random digits, the theoretical mean, or mean of the population, is 4.5 and the standard deviation is 2.872. These can be worked out by the methods of Lessons 2 and 4. The observed mean of the simulated sampling distribution is 4.364, quite close to what the theory predicts. The observed standard deviation of the simulated sampling distribution is 0.952, again quite close to $\frac{2.872}{\sqrt{10}} = 0.908$. The normal distribution seems to be a good model here.

STUDENT PAGE 48

Another Simulated Sampling Distribution

Do you think the normal distribution would apply to sampling distributions of means for other population distributions? After all, we have based most of our discussion on a single underlying distribution–that of the number of children in American families.

7. This simulation will involve the use of random digits. You will need a random-digit table or a calculator that generates random digits. The simulation is completed most efficiently by working in groups.

a. Work with your group to select 50 sets of ten random digits each from a random-number table, or random-number generator in your calculator. Compute the mean of each set of ten digits.

b. Plot the 50 means on a dotplot. Describe the shape. Does the plot look normal?

c. Compute the mean and the standard deviation of the set of 50 sample means.

d. Random digits take on integer values from 0 to 9 with equal probability. Using this fact, compute the expected value and the standard deviation as an expected value for the theoretical distribution of random numbers.

e. How does the observed mean of the simulated sampling distribution compare with the theoretical expected value? How does the observed standard deviation of the simulated sampling distribution compare with the theoretical standard deviation? Does it look as if the rules developed above for the sampling distribution for means apply in this case?

8. Consider the seeming generality of the normal model for sampling distributions of means, for a random sample of size n from a population with mean μ and standard deviation σ. Give the probability that the following inequality holds:

$$|\bar{x} - \mu| \leq \frac{2\sigma}{\sqrt{n}}$$

where $\bar{x}$ denotes the sample mean. (NOTE: This statement is the same as, "What is the probability that the sample mean will be within two standard deviations of its population mean?")

48 LESSON 6

8. The probability that the distance between the sample mean and the population mean is less than two standard deviations of the sampling distribution for the mean is approximately 0.95 because of the approximate normality of this sampling distribution.

STUDENT PAGE 49

Practice and Applications

9. **a.** $0.91 \pm 2\frac{1.114}{\sqrt{25}}$; 0.91 ± 0.45 or (0.46, 1.36)

b. $0.91 \pm 2\frac{1.114}{\sqrt{100}}$; 0.91 ± 0.22 or (0.69, 1.13)

c. $0.91 \pm 2\frac{1.114}{\sqrt{1000}}$; 0.91 ± 0.07 or (0.84, 0.98)

d. $0.91 \pm 2\frac{1.114}{\sqrt{4000}}$; 0.91 ± 0.04 or (0.87, 0.95)

10. The sample total = n·(sample mean) and an interval for sample totals can be found by simply multiplying both ends of the appropriate interval for means by the common sample size.

a. (11.5, 34), where $34 = 25 \cdot 1.36$ and $11.5 = 25 \cdot 0.46$

b. (69, 113), where $113 = 100 \cdot 1.13$ and $69 = 100 \cdot 0.69$

c. (840, 980), where $980 = 1000 \cdot 0.98$ and $840 = 1000 \cdot 0.84$

d. (3480, 3800), where $3800 = 4000 \cdot 0.95$ and $3480 = 4000 \cdot 0.87$

11. Because this is a difficult problem, we have provided the following steps which lead to its solution.

The point that cuts off the lower 0.025, or 2.5%, on the normal distribution is 2 standard deviations below the mean.

If the total number of children falls below 1000, then the mean number of children per family falls below $\frac{100}{n}$.

The mean of the sampling distribution for sample means is 0.91.

Therefore, we must choose the sample size n so that the point 1000/n lies 2 standard deviations

Practice and Applications

9. For the random samples of families that the Nielsen Company could select, what interval should contain the middle 95% of the sample means on number of children per family for samples of each size?

a. 25

b. 100

c. 1000

d. 4000, which is the approximate real sample size for Nielsen ratings

10. What interval should contain the middle 95% of possible values for the total number of children in random samples of each size?

a. 25

b. 100

c. 1000

d. 4000

11. How large should the sample size be to insure that the total number of children in the sample is below 1000 with probability of only about 0.025?

12. Below is a restatement of the question at the beginning of this lesson. Read the statement and approximate the chances asked about. Do you need to make any assumptions for your answers to be valid?

Suppose records show that the weekly amounts your school earns on recycling aluminum cans has a mean of $120 and a standard deviation of $8. In the sixteen-week period of the fall term, what is the chance that the mean weekly earnings will exceed $130? $124? $100?

13. Here is a restatement of Problem 8 from Lesson 5:

The people using an elevator in an office building have an average weight of approximately 150 pounds, and a standard deviation of weights of approximately 10 pounds. The elevator is designed for a 2000-pound weight maximum. This maximum can be exceeded on occasion, but should not be exceeded on a regular basis. Your job is to post a sign in the elevator stating the maximum number of people for safe use. Keep in mind that it is inefficient to make this

THE NORMAL DISTRIBUTION 49

below 0.91, or $\frac{1000}{n} - 0.91 = 2\frac{\sigma}{\sqrt{n}} = -2\frac{1.114}{\sqrt{n}}$.

Multiplying both sides by n and simplifying, we get the equation $0.91n - 2.228\sqrt{n} - 1000 = 0$ which can be solved by trial and error or by plotting the equation on a graphing calculator. The solution is $n = 1183$ or 1184.

12. For the sampling distribution of mean weekly earnings over a 16-week period, the standard deviation is $\frac{8}{4} = \$2$.

The $130 figure is 5 standard deviations above the mean, and has virtually no chance of being exceeded.

The $124 figure is 2 standard deviations above the mean and will be exceeded with probability 0.025.

The $100 figure is 10 standard deviations below the mean and will be exceeded with probability 1.

This analysis is based on the fact that the earnings for the 16 weeks of the term will behave like a random sample from the distribution of possible earnings. This may not be the case if the fall term in question is, for instance, unusually cold so that fewer cold drinks are consumed.

13. $n = 14$: The value $\frac{2000}{14} = 142.9$ is 2.68 standard deviations below the mean of 150. Thus, it will be exceeded between 97.50% and 99.85% of the time.

$n = 13$: The value $\frac{2000}{13} = 153.8$ is 1.37 standard deviations above the mean. Thus, it will be exceeded between 2.5% and 16% of the time, with the true percent being a little closer to the 16%.

$n = 12$: The value $\frac{2000}{12} = 166.7$ is 5.79 standard deviations above the mean, and should never be exceeded.

STUDENT PAGE 50

number too small, but dangerous to make it too large. What number would you use for maximum occupancy? Explain your reasoning.

What is the approximate probability that the weight limit will be exceeded if the number of people on the elevator is 14? 13? 12?

14. The normal distribution sometimes works well as a model of the distribution of the population itself. If the population has a normal distribution, then the sampling distribution for a sample mean will be normally distributed for any sample size, even a sample size of 1. The three histograms that follow show basketball scores from all NCAA college playoff games between 1939 and 1995. Plot A is based on the score of the winner, plot B on the score of the loser, and plot C on the total points scored in the game.

NCAA Basketball Scores

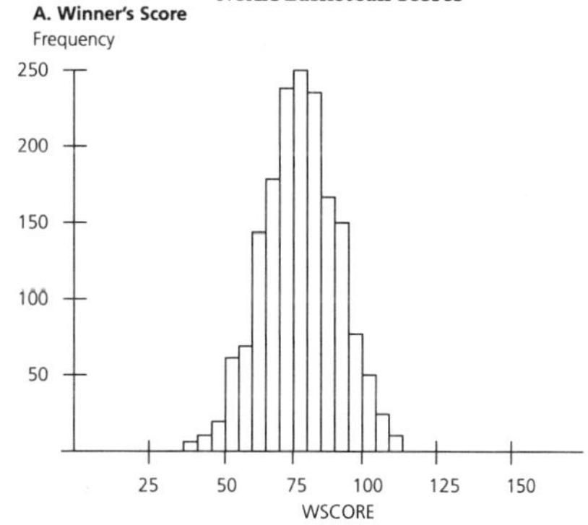

50 LESSON 6

14. a. Yes; an appropriate normal distribution would serve as an adequate model for all three distributions. The distribution of total scores is most nearly normal. The distribution of winner scores has a slightly extended tail on the side of the greater values, which is reasonable since there is a tendency for scores of the winners to be large. The distribution of loser scores has a slight bulge on the lower side of the mean; that is, it is not quite symmetric. This, too, is reasonable since the scores of the losers tend to be small, but not extremely so.

b. Winners' scores: mean, approximately 77; standard deviation, approximately 14

Losers' scores: mean, approximately 66; standard deviation, approximately 13

Total scores: mean, approximately 144; standard deviation, approximately 26

c. The interval 77 ± 2(14) or (49, 105) should include about 95% of the winners' scores. The bars on the histogram below 50 and above 105 indicate that this is approximately correct.

d. A total score of 180 is only 1.4 standard deviations above the mean. Thus, it could be exceeded fairly often, between 2.5% and 16% of the time. From the data that make up the histogram, the score of 180 was exceeded 130 times out of 1706 scores, or 7.6% of the time.

STUDENT PAGE 51

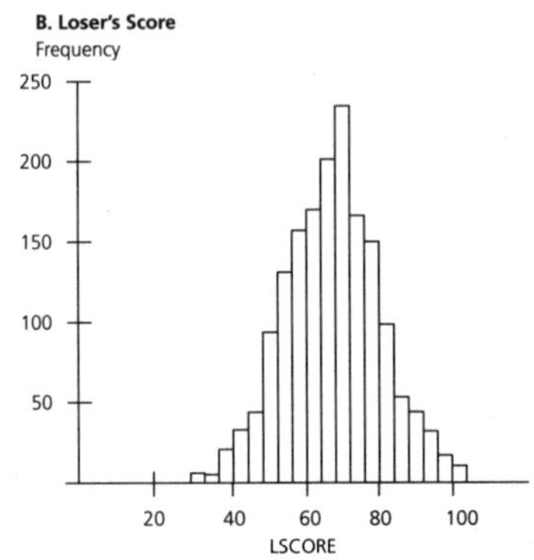

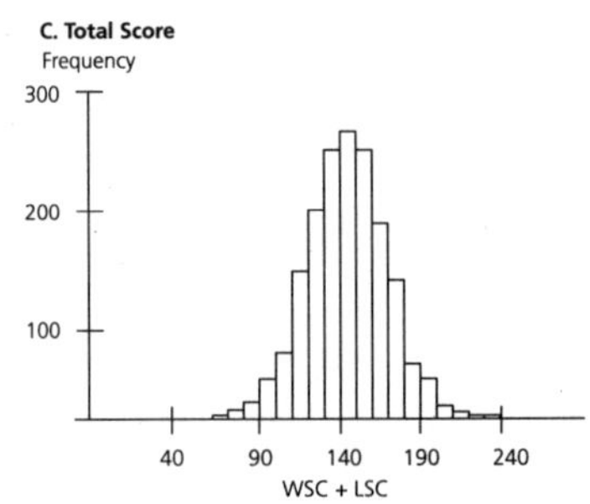

a. Does the normal distribution seem like a good model for these distributions? Which of the three do you think is least normal-looking? Explain.

b. Approximate the mean and standard deviation for each of the three distributions.

c. Find an interval which includes approximately 95% of the scores of the winning team.

d. Would you expect to see the total score in an NCAA playoff game go above 180 very often? Explain.

THE NORMAL DISTRIBUTION 51

STUDENT PAGE 52

SUMMARY

Sampling distributions for sample means, the distributions of potential values of a sample mean, are used widely in making decisions with data. The ***normal distribution*** provides a good ***model*** for these sampling distributions as long as the samples are randomly selected from a fixed population. The normal distribution is characterized by its symmetric mound shape, mean, and standard deviation. From knowledge of the mean and standard deviation, relative frequencies within intervals to either side of the mean can be found.

LESSON 7

The Distribution of a Sample Proportion

Materials: physical simulation device such as spinners or numbered chips for each group of students
Technology: graphing calculators or computers (optional)
Pacing: 1–2 class periods

Overview

Sample proportions in random sampling behave in a manner that exactly parallels the behavior of sample means, and this lesson has a structure that parallels that of Lesson 5. The sampling distribution of sample proportions is approximately normal with center at the true population proportion and with standard deviation that decreases as the sample size increases according to a known formula. This amazing and useful result allows us to make practical decisions on problems that involve sample proportions. It might be a good strategy to have students conduct the simulation and collect the data in one session and report on the results in another.

Teaching Notes

Again, the approximate normality of the sampling distribution for a proportion is best discovered by simulation. It is good if at least part of the simulation is done by hand so that students can see what is happening. A calculator or computer can be used to record results and graph the data, as in the figure that accompanies the lesson. The simulation requires an event with probability 0.6. This can be accomplished with physical devices such as spinners and numbered chips. Consider some of these options before going to technology. Make sure that students understand what is meant by a *proportion* in the sense of working with data.

Technology

As mentioned above, graphing calculators and computers can be used to summarize the data from the simulation, but it would be good if some of the simulation could be done with physical devices. It is quite easy, however, to do this simulation with random digits generated by a calculator or computer.

Follow-Up

The lesson outlines a generic simulation. Have students find an article in the media reporting on a sample survey with results summarized in a proportion. These surveys generally report a margin of error. Ask the students to design a simulation to see if the reported margin of error appears to be correct.

STUDENT PAGE 53

LESSON 7

The Distribution of a Sample Proportion

How many high-school graduates would you expect to see in a random sample of 1000 adults?

Is there a good chance that a sample of 1000 adults could produce over 900 high-school graduates?

Is the distribution of potential values of a sample proportion similar to the distribution of potential values for a sample mean?

What percent of the students in your school like the food in the cafeteria? What proportion of your income do you spend on food and entertainment? What fraction of the residents of your town own their homes? Data are often summarized by reporting a percent or a proportion for one or more categories involved. In order to make intelligent decisions based on data reported this way, we must understand the behavior of proportions that arise from random samples.

OBJECTIVES

Gain experience working with proportions as summaries of data.

Develop sampling distributions for sample proportions.

Discover the meaning of margin of error in surveys.

INVESTIGATE

The Sampling Distribution of a Proportion

First, let's be clear about the basic terminology. Suppose 20 students are asked to report their favorite food and 12 say "pizza." The *proportion* reporting pizza is $\frac{12}{20} = 0.60$ and the *percent* reporting pizza is 60%. Most of the problems in this lesson will deal with proportions, which will be numbers between 0 and 1.

Solution Key

Discussion and Practice

1. **a.** Most students will see that 40(0.6) = 24 is a reasonable answer.

b. Students are not expected to calculate an exact answer here. Some will reason that 30 does not seem far from the expected value of 24, so the probability might be fairly high at 0.3 or so.

c. For many students, 10 will seem like a long way from the expected value of 24, so the probability should be small.

2.–3. This simulation can be done quite easily with the random-number generator in a graphing calculator or computer. Statistical software packages often have routines built in to do this kind of random sampling. If you do not have access to technology of this type, random digits can be physically drawn out of a box.

The results given in Problem 4 demonstrate one possible set of outcomes for a simulation of this type.

STUDENT PAGE 54

Discussion and Practice

Thinking About Survey Results

1. It is estimated that about 60% of automobile drivers use seat belts. Suppose your class is to conduct a survey of 40 randomly selected drivers. Think about the following questions. Try to arrive at reasonable answers based upon your current knowledge and your intuition. Do not spend a lot of time with calculations.

a. How many drivers would you expect to be using seat belts?

b. What is the chance that more than 30 drivers will be using seat belts?

c. Would it be quite unusual to find fewer than 10 of the drivers wearing seat belts?

Now, we will develop the tools we need to answer these questions more precisely.

An intuitive "feel" for how sample percents behave in random sampling can be developed through simulation. Let Y denote the number of successes in a random sample of size n from a population in which the probability of success on any one sample selection is given by p. This section will concentrate on properties of the sample proportion Y/n. Y/n can be changed into a percent by multiplying by 100, but we will work most directly with the decimal form. This investigation can be completed most efficiently if you work in pairs.

Designing and Conducting the Simulation

2. Assume that $p = 0.6$ for this study.

a. Find a device that will generate an outcome that has probability equal to 0.6 of occurring. A random-number table, a calculator that generates random numbers, or slips of paper may be used.

b. Let the sample size be $n = 10$. Generate ten outcomes by the device agreed upon in Problem 2a and count the successes among the ten outcomes. Divide the number of successes by 10 to obtain the proportion of successes. That is the sample proportion that will be recorded.

c. Repeat Problem 2b three more times so that your group has four different samples of size 10 each and four observed values of the sample proportion.

54 LESSON 7

STUDENT PAGE 55

d. Combine your values of the sample proportion with those from the rest of the class. Plot the sample proportions on a dotplot or a stem-and-leaf plot.

e. Based on the results of the simulation, find approximate values for each probability.

i. $P(0.5 < \frac{Y}{n} < 0.7)$

ii. $P(0.4 < \frac{Y}{n} < 0.8)$

iii. $P(\frac{Y}{n} > 0.8)$

f. Calculate the mean and the standard deviation of the simulated distribution of sample proportions. Keep these for future reference.

g. From your dotplot or stem-and-leaf plot, describe the distribution of the possible values of $\frac{Y}{n}$ for samples of size 10, when the true value of p is 0.6. NOTE: This distribution is called a simulated sampling distribution of the proportion $\frac{Y}{n}$.

Changing the Sample Size

3. a. Repeat the simulation of Problem 2 for samples of size $n = 20$. Keep p fixed at 0.6. To save time, two of the size-10 samples may be combined to obtain a size-20 sample. Complete all parts of Problem 2.

b. Repeat the simulation of Problem 2 for samples of size $n = 40$. Keep p fixed at 0.6. To save time, two of the size-20 samples may be combined to obtain a size-40 sample. Complete all parts of Problem 2.

c. Compare the shapes of the simulated sampling distributions for samples of size 10, 20, and 40.

d. Compare the means of the simulated sampling distributions for samples of size 10, 20, and 40. How close are these means to the true value of p?

e. Compare the standard deviations of the simulated sampling distributions for samples of size 10, 20, and 40. A theoretical result in probability states that the standard deviation of sample proportions should be related to the true p and the sample size by the following rule:

$$SD(\text{proportion}) = \sqrt{\frac{p(1-p)}{n}}$$

STUDENT PAGE 56

Does this rule appear to hold for the sampling distributions generated in this investigation?

Summary of Sampling Distributions for Sample Proportions

4. Your plots may look something like those shown below. These were generated by computer for the same cases described above, with the addition of a sampling distribution for samples of size 80.

Sampling Distributions of Sample Proportions

Each dot represents 2 points.

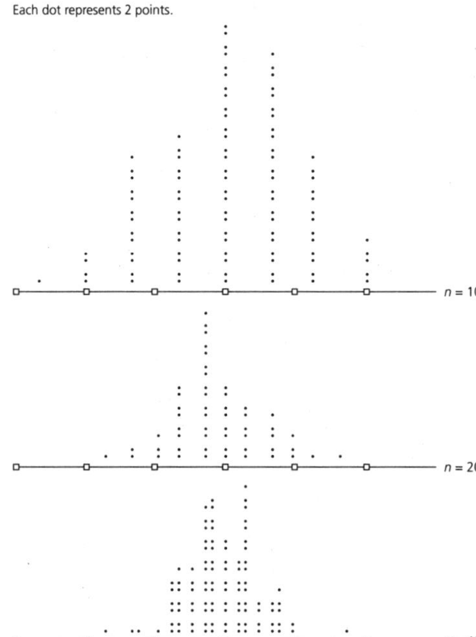

4. **a.** All four of the sampling distributions are somewhat mound-shaped and symmetric; at least they do not have a long tail in either direction. All four center at about the same place on the real-number line. The variation in the distributions decreases as the sample size increases.

b. All four of the distributions center close to 0.6, which is the population proportion of successes.

c. Using the rule *SD*(proportion) $= \sqrt{\frac{p(1-p)}{n}}$ with $p = 0.6$, the theoretical standard deviations are as follows:

$n = 10$: $SD = 0.1549$

$n = 20$: $SD = 0.1095$

$n = 40$: $SD = 0.0775$

$n = 80$: $SD = 0.0548$

These are quite close to the observed standard deviations of the sampling distributions shown in the lesson, which are 0.1552, 0.0962, 0.0838. and 0.0540, respectively.

STUDENT PAGE 57

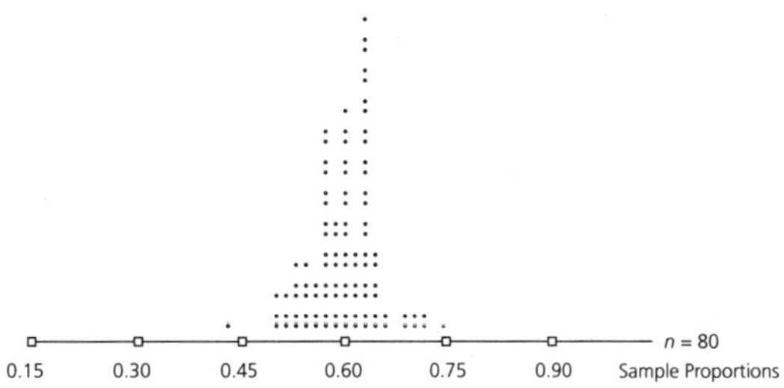

	Mean	Median	Standard Deviation
$n = 10$	0.6070	0.6000	0.1552
$n = 20$	0.5790	0.5500	0.0962
$n = 40$	0.5910	0.5750	0.0838
$n = 80$	0.5976	0.6000	0.0540

a. What are the main similarities and differences among these plots?

b. Where are the centers of these distributions? Are they all close to 0.6?

c. The observed standard deviations for the four sampling distributions are given following the four plots. How do these compare with the theoretical standard deviations calculated by the following rule?

$$SD(\text{proportion}) = \sqrt{\frac{p(1-p)}{n}}$$

Does the rule appear to work reasonably well?

The Normal Model for Proportions

From your simulation and the one seen in Problem 4, it appears that the sampling distributions for proportions are somewhat mound-shaped and symmetric. Thus, they could be modeled by the normal distribution with mean p and standard deviation

$$\sqrt{\frac{p(1-p)}{n}}$$

The following questions are intended to amplify this normality of the sampling distributions for proportions.

STUDENT PAGE 58

5. Refer to the four sampling distributions in Problem 4. Remember, in each case $p = 0.6$.

a. Find the fraction of the simulated values of $\frac{Y}{n}$ that were within two standard deviations of p. Do this separately for each of the distributions in the figure.

b. Symbolically, the fractions found in Problem 5a are estimates of

$$P\left[\left|\frac{Y}{n} - p\right| \le 2\sqrt{\frac{p(1-p)}{n}}\right]$$

For each of the cases $n = 10$, $n = 20$, $n = 40$, and $n = 80$, are these estimated probabilities close to each other?

c. Theoretically, about 95% of the observed values of a sample proportion will be within two standard deviations of the expected value of that proportion. This implies that it is quite likely, in any one sample, to obtain a sample proportion of successes that is less than two standard deviations from the "true" proportion of successes in the population. Do the results of Problem 5b appear to bear this out?

Practice and Applications

6. Look at the results of some recent polls as reported in the media—newspapers, news magazines, and so on.

a. Do the results of these polls tend to be reported in terms of counts, that is, number of "successes," or in terms of percents?

b. Discuss reasons why percents might be better than counts as a way to summarize data from polls.

c. Discuss aspects of the reporting on polls that you would like to see improved. Base the discussion on printed articles you have read.

7. If you look carefully at articles from the media on opinion polls, which are sample surveys, you may observe that many of them contain a statement similar to one of the following:

"The margin of error in these percents is 3%."

"The sampling error in any one reported percent is 2.5%."

What do you think they mean by these phrases? (HINT: Look again at Problem 5.)

5. **a.** $n = 10$: $0.60 \pm 2(0.1549) = 0.91$ or 0.29 contains 99/100 of the data values.

$n = 20$: $0.60 \pm 2(0.1095) = 0.82$ or 0.38 contains 98/100 of the data values.

$n = 40$: $0.60 \pm 2(0.0774) = 0.75$ or 0.44 contains 96/100 of the data values.

$n = 80$: $0.60 \pm 2(0.0547) = 0.71$ or 0.49 contains 94/100 of the data values.

The actual data values inside these intervals are a little difficult to discern from the graph, but a close approximation is all that is needed here.

b. Yes; the four fractions are relatively close to each other. The estimate of the probability in question should be the same for each of the four sample sizes.

c. Yes; all four of the estimates of the probability of a sample proportion being within two standard deviations of p are about 0.95, as the theory states.

Practice and Applications

6. **a.** Possible answer: Many newspapers and news magazines such as *Time, Newsweek,* and *U.S. News and World Report* almost always contain at least one poll. The results for categorical questions are always reported in terms of percents.

b. Possible answer: Percents adjust for the sample size, so it makes sense to compare results for two different polls or for two different groups within a poll by looking at percents. For example, your school could be compared with a much larger or much smaller school on the question of student participation in athletics by looking at the percents of students who compete in each school. It would not be fair to compare the actual numbers of students participating in athletics.

c. Possible answer: Often the details on how the sampling was done are not given. The exact questions asked of the participants may not be given. Some of the sample sizes and percents you would like to have for analysis may not be given in the report.

7. The "margin of error" or the "sampling error" is the quantity $2SD(\text{proportion}) = 2\sqrt{\frac{p(1-p)}{n}}$.

This tells you that the sample proportion is quite likely to lie within this distance of the true population proportion.

STUDENT PAGE 59

8. Speculate as to what would happen to the shapes, means, and standard deviations of the simulated sampling distributions of in $\frac{Y}{n}$ Problem 4 if p were changed to 0.5. What if p were changed to 0.8?

9. According to the U.S. Bureau of the Census, about 80% of U.S. residents over the age of 25 are high-school graduates. A Gallup poll is to be conducted among those over the age of 25—we'll call these "adults" for now—on issues surrounding education. So the Gallup organization wants to be sure there will be adequate numbers of both high- school graduates and non-graduates in the resulting random sample of adults.

a. What is the probability that a randomly selected adult is a high-school graduate?

b. If 1000 adults are randomly sampled, how many high-school graduates would you expect to see?

c. If Gallup expects to have 400 non-graduates in the sample, how many adults should be randomly selected?

d. Can Gallup be assured of at least 500 non-graduates in a poll of any fixed size?

e. If Gallup randomly samples 400 adults, is it likely that the poll will result in more than 335 high-school graduates? Explain.

10. Here is a restatement of the first set of questions posed in this lesson. Answer them as specifically as you can with the information gained in this lesson.

It is estimated that about 60% of automobile drivers use seat belts. Suppose your class is to conduct a survey of 40 randomly selected drivers. Think about the following questions. Try to arrive at reasonable answers based upon your current knowledge and your intuition. Do not spend a lot of time with calculations.

a. *How many drivers would you expect to be using seat belts?*

b. *What is the chance that more than 30 drivers will be using seat belts?*

c. *Would it be quite unusual to find fewer than 10 of the drivers wearing seat belts?*

8. Possible answer: If $p = 0.5$ rather than 0.6, the centers of the distributions would shift to 0.5 but the shapes would remain about the same. If $p = 0.8$ rather than 0.6, the centers of the distributions would shift to 0.8. The shapes would become more skewed toward the smaller values, especially for small sample sizes, since the proportions are constrained to be between 0 and 1.

9. **a.** $P(\text{high-school graduate}) = 0.8$

b. $1000(0.8) = 800$

c. If $n(0.2) = 400$, then $n = \frac{400}{0.2} = 2000$

d. Possible answer: No; the chance of seeing a non-graduate is small (0.2) and, therefore, a large sample would be required to have a good chance of seeing 500 non-graduates. But, because of the random nature of the sample, there is no guarantee that a sample of any size, no matter how large, would produce at least 500 non-graduates.

e. For $n = 400$ and $p = 0.8$, the standard deviation of the sampling distribution of sample proportions is 0.02. The sample proportion of $\frac{335}{400} = 0.84$ is 2 standard deviations above the mean of the sampling distribution. Thus, the chance of seeing more than 335 graduates in a sample of 400 is about 0.025.

10. **a.** The expected value is $40(0.6) = 24$.

b. For samples of size 40 and $p = 0.6$, the standard deviation of the sampling distribution of sample proportions is 0.0774. The sample proportion of $\frac{30}{40} = 0.75$ is 1.94 (nearly 2) standard deviations above the mean of the distribution and, therefore, the chance of seeing more than 30 drivers using seat belts is a little over 0.025.

c. With a standard deviation of 0.0774, a sample proportion of $\frac{10}{40} = 0.25$ is about 4.5 standard deviations below the mean. The chance of seeing a sample proportion less than 0.25 is essentially 0. It would be quite unusual to see such a result in a random sample of 40 drivers.

STUDENT PAGE 60

SUMMARY

Sample ***proportions*** or percents are used often to summarize data on categorical variables, like gender, ethnic group, attitude on public issues, and health condition. In order to make decisions based on this kind of data, we must know something about the anticipated behavior of sample proportions from random samples. What we have discovered is that sample proportions have approximately normal sampling distributions, with mean equal to the true population proportion p for the characteristic being studied and standard deviation given by the following rule:

$$\text{SD(proportion)} = \sqrt{\frac{p(1-p)}{n}}$$

ASSESSMENT

Lessons 5 – 7

STUDENT PAGE 61

ASSESSMENT

Lessons 5–7

1. Explain what a sampling distribution for a sample mean is and how it differs from the underlying population distribution.
2. According to the U.S. Bureau of the Census, the number of people per household averages 2.58 and the standard deviation of the number of people per household is 1.39. (NOTE: See Lesson 2 for the actual population distribution for this variable.) Describe the sampling distribution of the sample mean for random samples from this distribution for each sample size.
   a. $n = 5$
   b. $n = 100$
   c. $n = 1000$

   Your description should include sketches of the sampling distributions.
3. Based on the information in Problem 2 on the distribution of household sizes, how large a sample size is needed to guarantee that the standard deviation of the distribution of possible sample means is less than 0.10?
4. Gallup is conducting a poll of American households. A typical sample size is 1200 households. For a sample of 1200 households, what interval should contain the middle 68% of the possible values for the sample mean?
5. What sample size should Gallup use if it is desired to have a total of at least 2000 people in the sample with probability about 0.975?
6. Suppose the distribution of household size in the U.S. has summary statistics as given in Problem 2. For a sample of 1200 households, the distance between the sample mean and the population mean should still be less than K with probability about 0.95. Find K. Did you make any assumptions in the process of finding K? What are they?

LESSONS 5–7 61

Solution Key

1. A sampling distribution for a sample mean shows the distribution of the possible values of the sample mean that could occur in a random sample from a fixed population. The sampling distribution is the distribution of potential values of the mean that could be produced by a random sample, thus giving the investigator some idea of what might happen when a random sample is actually taken.

   The underlying population distribution shows the distribution of the data values in a population before any sampling is done. A random sample of measurements is selected from the population distribution to produce the mean on which the sampling distribution is based.

   Sampling distributions depend upon random sampling and describe potential values of a statistic. Population distributions describe the actual values a variable can take on.

2. **a.** $n = 5$: the sampling distribution of the sample mean is approximately normally distributed with mean 2.58 and standard deviation 0.62, although for samples this small the sampling distribution may retain some of the features of the population distribution.

   b. $n = 100$: the sampling distribution of the sample mean is approximately normally distributed with mean 2.58 and standard deviation 0.14.

   c. $n = 1000$: The sampling distribution of the sample mean is approximately normally distributed with mean 2.58 and standard deviation 0.04.

   The sampling distributions will look more and more normal as the sample size increases.

3. To achieve this goal, $\frac{\sigma}{\sqrt{n}} = \frac{1.39}{\sqrt{n}}$
   < 0.10
   or $\sqrt{n} > 13.90$
   or $n > 193.21$ or 194.

4. The interval $\mu \pm \frac{\sigma}{\sqrt{n}}$ or $2.58 \pm \frac{1.39}{\sqrt{1200}}$ or 2.58 ± 0.04 yields the interval (2.54, 2.62), which means that the interval should contain the middle 68% of possible values of the sample mean.

5. If the total is to be at least 2000, then the sample mean must be at least $\frac{2000}{n}$, where n is the sample size. The upper 97.5% of the normal distribution falls above a point which is 2 standard deviations below the mean of the sampling distribution, 2.58 in this case. Thus, we have the equation
   $$\frac{2000}{n} - 2.58 = -2\frac{1.39}{\sqrt{n}}$$
   or $2.58n - 2.78\sqrt{n} - 2000 = 0$.

This equation can be solved by trial and error or with a graphing calculator or computer to yield $n \approx 806$.

6. The distance between the sample mean and the population mean is less than 2 standard deviations of the sampling distribution with probability 0.95. Therefore, K must equal 2 standard deviations, or $K = 2\frac{\sigma}{\sqrt{n}} = 2\frac{1.39}{\sqrt{1200}} = 2(0.04) = 0.08$. The assumption is that the 1200 households must be selected randomly from the population of U.S. households.

7. The distribution of possible values of the sample proportion is approximately normal in shape, with center at p and with standard deviation given by $\sqrt{\frac{p(1-p)}{n}}$.

8. The interval containing the middle 95% of the potential sample proportions is

$$p \pm 2\sqrt{\frac{p(1-p)}{n}} =$$

$$0.6 \pm 2\sqrt{\frac{0.6(0.4)}{100}}$$

$= 0.6 \pm 2(0.049)$
or (0.502, 0.698)

9. The margin of error is 2 standard deviations of the sampling distribution. Thus, the sample size n must satisfy $2\sqrt{\frac{0.6(0.4)}{n}} = 0.04$

$$\sqrt{n} = \frac{2\sqrt{0.6(0.4)}}{0.04}$$

or $n = \frac{4(0.6)(0.4)}{(0.04)^2} = 600$.

10. a. With $p = 0.3$ and $n = 500$, the standard deviation of the sampling distribution is 0.02. A sample proportion of 0.4 is 5 standard deviations above the mean of the sampling distribution, and so it would be very unusual to see such a result.

b. This would be a very unusual result under the usual conditions of good sampling practice. There are three possible explanations for such a result:

The sample may not have been selected randomly.

The population figure of 30% without health insurance could be wrong.

Everything was done properly and the population figure is correct, but the investigator obtained a result that would very rarely occur.

STUDENT PAGE 62

7. A random sample of n students in your school is to be selected to estimate the proportion p of students who will be requesting parking spaces for next year. Describe the shape, center and spread of the distribution of possible values of the sample proportion that could result from this survey.

8. About 60% of drivers wear seat belts. In a random sample of 100 drivers, what interval should contain the middle 95% of the possible sample proportions of seat belt users?

9. From prior studies we assume that the percent of drivers wearing seat belts is around 60%, but we do not know for sure. A new survey is commissioned to study this issue. What sample size will guarantee a margin of error no larger than 0.04?

10. About 30% of the residents of the U.S. are without health insurance.

a. If a poll of 500 residents is conducted, with random sampling, would it be unusual to find over 40% without health insurance? Explain.

b. In a poll of 500 residents, suppose 42% were found to be without health insurance. Discuss various possible reasons for this unusually large sample percent.

11. The plots on the next page show the percent of U.S. adult females and adult males having heights that would round off to the integer values given on the horizontal axes. The female percents are given for heights from 56 through 71 inches and the male percents are given for heights from 61 through 76 inches.

62 ASSESSMENT

11. **a.** The female heights have a mean of about 65 inches and a standard deviation of about 2 inches. The mean is nearly at the center of this somewhat symmetric distribution, but is pulled a little lower by the long tail to the left. The standard deviation is half the thickness of the curve about $\frac{2}{3}$ of the way up the center line, the line of symmetry.

b. The male heights have a mean of about 70 inches and a standard deviation of a little more than 2 inches, say about 2.5 inches. The reasoning is the same as in Problem 11a.

c. A height 1 standard deviation above the mean will satisfy this condition, which gives 65 + 2 = 67 inches.

d. A height one standard deviation below the mean will satisfy this condition, which gives 70 – 2.5 = 67.5 inches.

STUDENT PAGE 63

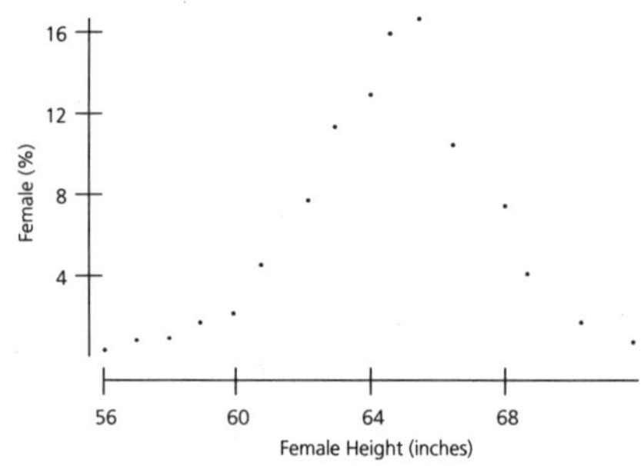

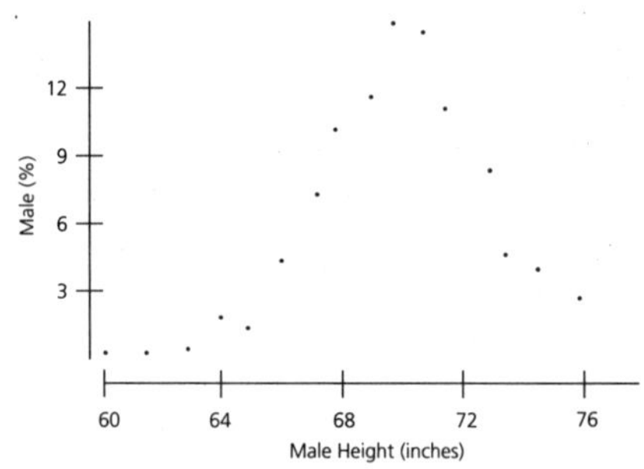

a. What are the approximate mean and standard deviation of female heights?

b. What are the approximate mean and standard deviation of male heights?

c. Find a female height that would be exceeded by only about 16% of the adult females in the U.S.

d. Find a male height that has probability 0.84 of being exceeded by the height of a randomly select adult male from the U.S.

UNIT III

Two Useful Distributions

LESSON 8

The Binomial Distribution

Materials: none
Technology: graphing calculators or computers (optional)
Pacing: 2 class periods, with exercises assigned as homework

Overview

The approximate normality of the distribution of sample proportions works well only when the sample size is reasonably large. In the case of small samples, we must use a more precise result for the probability of the number of "successes" in a random sample of size n. The probabilities that result from a simple model for this situation turn out to be the terms in a binomial expansion.

Teaching Notes

This is the most mathematical of the lessons so far. Students may need to be reminded of some of the basic rules of probability, such as the multiplication rule for independent events and the addition rule for mutually exclusive events. It is important for them to see how the mathematics of probability comes together to build a very useful model.

Students can profit from group work in the development of the binomial model, but should work through some of the exercises on their own.

Technology

Some of the binomial probabilities, for small sample sizes, should be calculated by hand so that students can see each component part. After that experience, binomial probabilities can be calculated automatically on most graphing calculators or statistics software packages.

Follow-Up

Have students compare the exact binomial probabilities they get from their calculators with the normal approximations from earlier lessons. Have them try this for sample sizes of 5, 10, 50, and 100 to see that the normal approximation is good for large samples but perhaps not so good for small ones. Note that the normal approximation is stated for proportions, so that students will have to translate from proportions to counts.

STUDENT PAGE 67

LESSON 8

The Binomial Distribution

In a random sample of four adults, what is the probability of sampling three high-school graduates?

How do small sample distributions differ from large sample distributions?

Of what use are mathematical models for calculating probabilities?

You learned in Lesson 7 that sample proportions tend to have normal sampling distributions; so the normal probability model works well for approximating probabilities associated with sample proportions. However, the normal approximations work well only in cases for which the sample size is large and the population proportion p is not close to zero or 1. For small samples or for cases in which p is far from 0.5, we need a more precise model.

OBJECTIVES

Understand the basic properties of the binomial distribution.

Use the binomial distribution as a model for certain types of counts.

INVESTIGATE

Sampling Distributions

It is reported that about 80% of adults over 25 in the U.S. are high-school graduates. Suppose a Gallup poll randomly selects only four adults from a small city for which this population percent is thought to hold. The first plot that follows shows a simulated sampling distribution for the sample proportion of high-school graduates in such a sample. The second plot shows the simulated distribution of the actual count of the number of high school graduates per sample of size 4.

Solution Key

Discussion and Practice

1. These distributions are highly skewed toward the smaller values. They look nothing like normal distributions, and could not be approximated well by normal distributions.

STUDENT PAGE 68

Sampling Distributions for a Small Sample

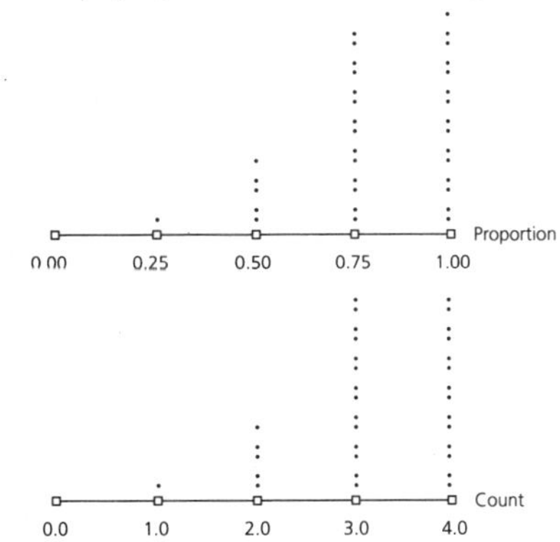

Discussion and Practice

1. Describe the shapes of the two distributions plotted. Do they look as if they could be approximated well by normal distributions?

We will not attempt to give a general rule as to when the normal distribution is an appropriate approximation to the sampling distribution for a proportion; when in doubt, set up a small simulation to see if you think the normal model will work well.

A Random Variable for a Categorical Outcome

In the Gallup poll mentioned above, the outcome for any one sampled person is either "high-school graduate" or "not a high-school graduate." What can we do with these categorical outcomes to change them into numerical ones? One way to make the transition from categorical to numerical is to assign a 1 if we see the characteristic we are looking for and to assign a 0 if we do not see that characteristic. We might, for example, define "high-school graduate" to be of primary interest in the Gallup poll. Thus, for the first person sampled, we would record a 1 if that person is a high-school graduate and a 0 if not. Symbolically, we could write $X_1 = 1$ if sample person 1 is a

68 LESSON 8

2. **a.** $E(X_1) = 0(1 - p) + 1(p) = p$

b. It is easier to work with the variance, the square of the standard deviation, at first. Let $V(X)$ denote the variance of a random variable X. Then,

$V(X_1) = E(X_1 - p)^2$
$= (0 - p)^2(1 - p) + (1 - p)^2(p)$
$= (p^2)(1 - p) + (1 - p)^2(p)$
$= (1 - p)(p^2 + (1 - p)(p))$
$= (1 - p)\,p = p(1 - p)$

and $SD(X_1) = \sqrt{p(1-p)}$

3. The probability distribution for X_2 is exactly like that for X_1.

X_2	$P(X_2)$
0	$1 - p$
1	p

STUDENT PAGE 69

high-school graduate and $X_1 = 0$ if sample person 1 is not a high-school graduate.

The random variable X_1, then, completely describes the outcomes of interest for the first person sampled. The probability distribution for X_1 can be written as

X_1	$P(X_1)$
0	$1 - p$
1	p

where p represents the probability of selecting an item with the characteristic of interest. In the high school graduate example, $p = 0.8$.

2. Use the random variable X_1 described above.

a. Find the expected value of X_1. How does it relate to the probability of sampling a high-school graduate?

b. Find the standard deviation of X_1. Write this as a function of p.

3. Suppose X_2 represents the outcome for a second adult sampled from the city, coded in the same way as X_1. Give the probability distribution for X_2. (HINT: The second adult is selected independently from the first.)

The Binomial Distribution

We are ready to develop a formula for the probability of obtaining a number Y of high-school graduates in a sample of n adults from a city in which p is the proportion of adults which are, in fact, high-school graduates. We begin by letting $n = 2$. In this case, Y can equal 0, 1, or 2; there are no other possibilities. Y is 0 if neither of the selected adults is a high-school graduate. Symbolically,

$$P(Y = 0) = P(X_1 = 0 \text{ and } X_2 = 0) = P(X_1 = 0) \cdot P(X_2 = 0) = (1 - p)^2.$$

Multiplication is valid here since you are looking at the intersection of two independent events. For the sake of convenience in writing, let $(1 - p) = q$. Y is 1 if exactly one of the adults is a high-school graduate. Symbolically,

$$\begin{aligned} P(Y = 1) &= P((X_1 = 1 \text{ and } X_2 = 0) \text{ or } (X_1 = 0 \text{ and } X_2 = 1)) \\ &= P(X_1 = 1 \text{ and } X_2 = 0) + P(X_1 = 0 \text{ and } X_2 = 1) \\ &= P(X_1 = 1) \cdot P(X_2 = 0) + P(X_1 = 0) \cdot P(X_2 = 1) \\ &= pq + qp = 2pq. \end{aligned}$$

4. Letting Y denote the total number of successes in a random sample of three adults:

$P(Y = 0) = P(X_1 = 0 \text{ and } X_2 = 0 \text{ and } X_3 = 0)$
$= P(X_1 = 0) \cdot P(X_2 = 0) \cdot P(X_3 = 0)$
$= (1 - p)^3 = q^3$

$P(Y = 1) = P[(X_1 = 1 \text{ and } X_2 = 0 \text{ and } X_3 = 0) \text{ or } (X_1 = 0 \text{ and } X_2 = 1 \text{ and } X_3 = 0) \text{ or } (X_1 = 0 \text{ and } X_2 = 0 \text{ and } X_3 = 1)]$
$= P[(X_1 = 1 \text{ and } X_2 = 0 \text{ and } X_3 = 0)] + P[(X_1 = 0 \text{ and } X_2 = 1 \text{ and } X_3 = 0)] + P[(X_1 = 0 \text{ and } X_2 = 0 \text{ and } X_3 = 1)]$
$= p \cdot q \cdot q + q \cdot p \cdot q + q \cdot q \cdot p$
$= 3p(q)^2$

$P(Y = 2) = P[(X_1 = 1 \text{ and } X_2 = 1 \text{ and } X_3 = 0) \text{ or } (X_1 = 1 \text{ and } X_2 = 0 \text{ and } X_3 = 1) \text{ or } (X_1 = 0 \text{ and } X_2 = 1 \text{ and } X_3 = 1)]$
$= p \cdot p \cdot q + p \cdot q \cdot p + q \cdot p \cdot p$
$= 3(p)^2q$

$P(Y = 3) = P(X_1 = 1 \text{ and } X_2 = 1 \text{ and } X_3 = 1) = p^3$

5. $E(Y)$
$= 0(q^3) + 1(3pq^2) + 2(3p^2q) + 3(p^3)$
$= 3p(q^2 + 2pq + p^2) = 3p(q + p)^2$
$= 3p$ since $(q + p) = 1$.

6. The expected value of 3.2 indicates the average number of high-school graduates that would appear in many random samples of size 4. Suppose 100 different polls each randomly selected four adults from the population. These 100 samples should average 3.2 high-school graduates per sample, for a total of 3.2(100) = 320 high-school graduates in all polls combined.

STUDENT PAGE 70

Why can we add probabilities on the second line?

Finally, Y is 2 if both adults are high-school graduates. Symbolically,

$$P(Y = 2) = P(X_1 = 1 \text{ and } X_2 = 1) = p^2.$$

By combining the above results, you can represent the probability distribution for Y in the case $n = 2$ with the following table.

Y	$P(Y)$
0	q^2
1	$2pq$
2	p^2

4. Show that the probability distribution of Y for the sample size $n = 3$ is given by the following table.

Y	$P(Y)$
0	q^3
1	$3pq^2$
2	$3p^2q$
3	p^3

From the probability distributions, we can find general expressions for the expected value and standard deviation of Y. We will look at the specific cases for $n = 1$, 2, and 3 and see what generalization this suggests. For the expected value, or mean, of the probability distribution, we have:

$n = 1 \quad E(Y) = E(X_1) = 0q + 1p = p$
$n = 2 \quad E(Y) = 0(q^2) + 1(2pq) + 2(p^2) = 2p(q + p) = 2p$

5. Using the distribution developed in Problem 4, show that $E(Y) = 3p$ when $n = 3$.

The obvious guess for a general result is that $E(Y) = np$ for distributions of this type, and that is correct. Thus, in a random sample of four adults from the city being studied, we would expect to see about 4(0.8) = 3.2 high-school graduates.

6. The expected number of high-school graduates is not an integer. Provide a meaningful interpretation of this number.

When developing expressions for standard deviations, it is easier to begin with the variance, the square of the standard deviation, and then take a square root at the end. We denote variance by V, and recall that the variance is the expected value

70 LESSON 8

7. As in Problem 2, it is easier to begin with the variance.

$V(Y) = E(Y - 2p)^2$
$= (0 - 2p)^2(q^2) + (1 - 2p)^2(2pq) + (2 - 2p)^2(p^2)$
$= 4p^2(q^2) + (1 - 2p)^2(2pq) + 4q^2(p^2)$
$= 2pq[4pq + (1 - 2p)^2]$
$= 2pq[4pq + (1 - 4p + 4p^2)]$
$= 2pq[4p(1 - p) + (1 - 4p + 4p^2)]$
$= 2pq(4p - 4p^2 + 1 - 4p + 4p^2)$
$= 2pq(1)$
$= 2pq$

8. Using the formula $SD(Y) = \sqrt{np(1-p)}$ with $p = 0.8$, it follows that the standard deviations for the respective sample sizes are as follows.

$n = 4$: 0.80
$n = 10$: 1.26
$n = 20$: 1.79
$n = 40$: 2.53

The standard deviations of the total number of high-school graduates in the samples increase as the sample size increases. This is quite different from the behavior of standard deviations of sample proportions, in which the standard deviation decreases as the sample size increases. This is why it is easier to accurately estimate a proportion than a total, based on random samples.

9. These are simple binomial expansions. If multiplication of algebraic expressions such as these has not been covered elsewhere, this question could take a little time to develop.

STUDENT PAGE 71

of the square of the deviations from the expected value. Using the probability distribution for X_1, which has expected value p, the variance becomes

$$V(X_1) = E(X_1 - p)^2 = (0 - p)^2q + (1 - p)^2p = p^2q + q^2p$$
$$= pq(p + q) = pq$$

and the standard deviation is

$$SD(X_1) = \sqrt{V(X_1)} = \sqrt{pq}$$

7. For $n = 2$, show that $SD(Y) = \sqrt{2pq}$.

The above results suggest the correct generalization for the standard deviation, which is

$$SD(Y) = \sqrt{npq}$$

8. Find the standard deviation you would expect to see among the number of high-school graduates observed in samples of size 4. Repeat for samples of size 10, 20, and 40. What do you notice about the standard deviations for the counts Y that differs from the standard deviations for proportions studied in Lesson 7?

Look back at the probability distributions for $n = 1$ and $n = 2$, $n = 3$, the one developed in Problem 4. You can begin to generalize to a formula that works for any sample size and see why the distribution is called the ***binomial distribution***. The sum of the probabilities across all possible values in a probability distribution must equal 1. Exploring this idea for the various sample sizes leads to

$n = 1$ $\quad q + p = 1$
$n = 2$ $\quad q^2 + 2pq + p^2 = (q + p)^2 = 1$
$n = 3$ $\quad q^3 + 3pq^2 + 3p^2q + p^3 = (q + p)^3 = 1$

9. For the case $n = 2$, expand the binomial expansion $(q + p)^2$ to show that it equals the expression on the left. Do the same for the case $n = 3$.

The interesting result is that all of the probabilities for each value of n are terms of the binomial expansion of $(q + p)^n$. That is the reason for the name binomial probability distribution. It provides a way of writing a general probability statement for any sample size.

For a random sample of size n from a population with probability of "success" p on each selection, the probability that the

THE BINOMIAL DISTRIBUTION 71

10. We show the answer for the case $n = 3$. The case $n = 2$ is a simpler version. For $n = 3$, c can take on the values 0, 1, 2, and 3.

$P(Y = 0) = \frac{3!}{0!3!} p^0(1 - p)^3$

$= (1 - p)^3 = q^3$

since $0! = 1$ by definition.

$P(Y = 1) = \frac{3!}{1!2!} p^1(1 - p)^2$

$= 3p(1 - p)^2 = 3pq^2$

$P(Y = 2) = \frac{3!}{2!1!} p^2(1 - p)^1$

$= 3p^2(1 - p)^1 = 3p^2q$

$P(Y = 3) = \frac{3!}{3!0!} p^3(1 - p)^0$

$= p^3$

Note that $c! = c(c - 1)(c - 2) \ldots (1)$ and $0! = 1$, which may require some explanation if students do not recall this fact.

Practice and Applications

11. a. The binomial probability distribution for the case $n = 4$ and $p = 0.8$ is as follows:

Y	P(Y)
0	0.0016
1	0.0256
2	0.1536
3	0.4096
4	0.4096

Students should be encouraged to use a calculator or computer to obtain these results.

b.

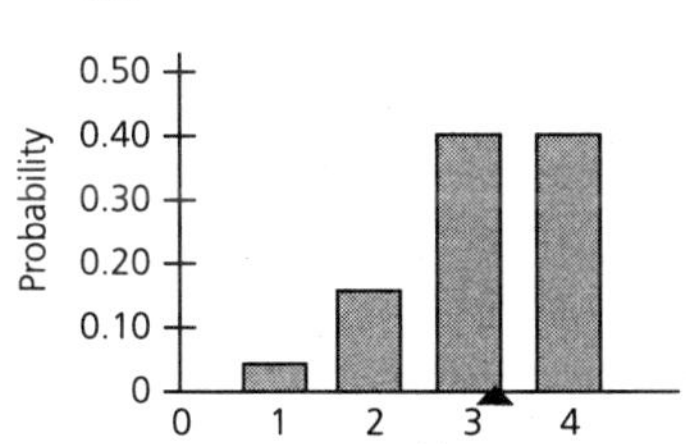

STUDENT PAGE 72

number of "successes" in the sample Y equals a specific count c is given by

$$P(Y = c) = \binom{n}{c} p^c (1 - p)^{n-c}$$

where $\binom{n}{c}$ is the binomial coefficient calculated as $\binom{n}{c} = \frac{n!}{c!(n - c)}$.

In this calculation, $n! = n(n - 1)(n - 2) \ldots 1$. This expression is called "n factorial."

10. Show that the formula for $P(Y = c)$ gives the expressions shown on the tables provided earlier in this lesson for the cases $n = 2$ and $n = 3$.

Practice and Applications

11. Here is a statement from the first page of this lesson.

It is reported that about 80% of adults over 25 in the U.S. are high-school graduates. Suppose a Gallup poll randomly selects only FOUR adults from a small city for which this population percent is thought to hold.

a. Write out the probability distribution for Y, the number of high-school graduates to be seen in a random sample of size 4. Use your calculator to find the numerical values.

b. Construct a bar graph of the probability distribution found in part a.

c. Find the expected value of Y. Mark the expected value of Y on the bar graph.

d. Find the standard deviation of Y.

e. Mark the point $E(Y)$ plus one standard deviation of Y on the bar graph. Similarly, mark the point $E(Y)$ minus one standard deviation of Y. What possible values of Y lie between these two points? What is the total probability for the values of Y between these two points?

12. A student is guessing at all five true-false questions on a quiz.

a. Give the probability distribution for the number of true-false questions the student gets correct.

b. If the student needs to answer at least 60% of the true-false questions in order to pass the quiz, what is the probability that the student will pass?

72 LESSON 8

c. $E(Y) = np = 4(0.8) = 3.2$

d. $SD(Y) = \sqrt{4(0.8)(0.2)} = 0.80$

e. The interval mean plus and minus 1 standard deviation covers the interval (2.4, 4.0). The values of Y that lie within this interval are 3 and 4, and the probability that either of these values will occur is $2(0.4096) = 0.8192$. Note that this probability is much greater than the probability of being within 1 standard deviation of the mean under the normal distribution.

12. a. In this case, the sample size is 5 and the probability of success on any one question is 0.5, since the student is guessing among two possible answers. The probability distribution of Y, the number of correct answers, is.

(12 a).

Y	$P(Y)$
0	0.03125
1	0.15625
2	0.31250
3	0.31250
4	0.15625
5	0.03125

b. Getting at least 60% of the questions correct implies that the student must guess three or more answers correctly. $P(Y \geq 3) = 0.50$, the sum of the last three entries of the table. This answer can be seen by realizing that this distribution is symmetric.

13. For a ten-question true-false test, with a student guessing on every answer, the probability distribution for Y, the number of correct answers, is:

Y	$P(Y)$
0	0.000977
1	0.009766
2	0.043945
3	0.117188
4	0.205078
5	0.246094
6	0.205078
7	0.117188
8	0.043945
9	0.009766
10	0.000977

A student must guess 6 or more answers correct in order to get at least 60% on this test. The probability of 6 or more correct answers is only 0.377, the sum of the last five entries on the table. The student has a better chance of obtaining a passing grade by guessing on the test with the fewer questions.

STUDENT PAGE 73

13. You are to take a true-false quiz for which you have not studied, so you must guess at all of the answers. You need to have at least 60% of the answers correct in order to pass. Would you rather have a 5-question quiz or a 10-question quiz? Explain your reasoning.

14. A blood bank knows that only about 10% of its regular donors have type-B blood.

a. Ten donors will appear at the blood bank today. What is the chance that the blood bank will get at least one donor with type-B blood?

b. One hundred blood donors will appear at the blood bank this month. What is the approximate probability that at least 10% of them will have type-B blood? Explain how you made this approximation.

c. The blood bank needs 16 type-B donors this month. If 100 donors appear this month, does the blood bank have a good chance of getting the amount of type-B blood it needs? What recommendation would you have for the blood-bank managers?

15. If Y represents the number of successes in a binomial distribution of n trials, then *Y/n* represents the proportion of successes in the *n* trials. Show that

$$E\left(\frac{Y}{n}\right) = p$$

and

$$SD\left(\frac{Y}{n}\right) = \sqrt{\frac{p(1-p)}{n}}$$

Is this consistent with what you learned in Lesson 7?

16. The median annual household income for U.S. households is about $39,000.

a. Among five randomly selected households, find the probability that four or more have incomes exceeding $39,000 per year.

b. Consider a random sample of 16 households.

i. What is the expected number of households with annual income below $39,000?

ii. What is the standard deviation of the number of households with annual income below $39,000?

14. a. The chance of at least one donor out of ten with type-B blood is $1 - (0.9)^{10} = 1 - 0.349 = 0.651$. The complement of "at least one" is "none." This can be solved directly with a calculator, but it can also serve as a good exercise in the use of logarithms.

b. Two ways of finding an answer to this problem have been presented in this module. The direct way is to use the binomial distribution with $n = 100$ and $p = 0.1$. A graphing calculator or computer can handle sample sizes this large. The exact calculation gives $P(Y \geq 10) = 0.549$.

An alternative is to use the normal approximation to the distribution of proportions. The sample proportion of donors with type-B blood in a random sample of size 100 has approximately a normal distribution with a mean of 0.1. Since the normal distribution is symmetric with respect to its mean, the chance of seeing a sample proportion larger than 0.1 is approximately 0.5.

c. As in Question 14b, there are two approaches to this problem. From direct binomial calculations, $P(Y \geq 16) = 0.04$.

Using the normal approximation for sample proportions, as outlined in Question 14b, requires the standard deviation, 0.03. Thus, 0.16 is exactly 2 standard deviations above the mean of the distribution, which is 0.10. Under the normal distribution, a randomly selected value lies more than 2 standard deviations above the mean with probability 0.025. This is not quite the same as the exact binomial value, but it is not a bad approximation to use in situations in which the technology to calculate the exact value may not be readily available. In either case, the blood bank does not have a good chance of getting the 16 type-B donors out of a group of 100 donors.

15. This result is a simple application of the fact that $E(cY) = cE(Y)$ and $SD(cY) = cSD(Y)$ for any positive constant c. These ideas were developed in earlier lessons.

16. a. Since the median income has half of the incomes below it and half above it, the chance of a randomly selected income exceeding the median is 0.5. So, the basis for the probability in question is a binomial distribution with $n = 5$ and $p = 0.5$. Letting Y denote the number of households with income exceeding the median, the probability distribution is as given above, right:

Y	$P(Y)$
0	0.03125
1	0.15625
2	0.31250
3	0.31250
4	0.15625
5	0.03125

The probability that four or more households have incomes exceeding the median, among five randomly sampled households, is 0.1875, the sum of the lowest two entries on the table. NOTE: It is good to have students get into the habit of looking at the whole distribution when answering specific probability questions of this type.

b. i. If X denotes the number of households in a random sample of 16 with incomes below the median, then X has a binomial distribution with $n = 16$ and $p = 0.5$. Thus, $E(X) = np = 16(0.5) = 8$.

ii. The standard deviation of X is given by $SD(X) = \sqrt{np(1-p)}$ $= \sqrt{16(0.5)(0.5)} = 2$.

iii. A quick, approximate answer for the probability of seeing at least 10 of the 16 households with incomes under the median comes by observing that 10 is exactly 1 standard deviation above the mean of 8. With $p = 0.5$, the binomial distribution will be mound-shaped and symmetric, so that normal approximations should be good. The probability of seeing a data value more 1 standard deviation above the mean under the normal distribution is 0.16.

The exact binomial calculation gives $P(X \geq 10) = 0.227$. Notice that the normal approximation did not work as well for $n = 16$ as it did for $n = 100$.

c. This would be a very rare result, and you might suspect that the sample was not selected randomly. The sample selection method seems biased toward households with larger incomes.

STUDENT PAGE 74

17. a. P(at least one alarm sounds)
$= 1 - P$(all alarms fail)
$= 1 - (0.3)^3 = 0.973$

b. $1 - (0.3)^6 = 0.99927$; no; the two probabilities are nearly the same.

c. $1 - (0.3)^n = 0.99$ has as its closest solution $n = 4$. This equation can be solved by trial and error or by logarithms.

18. a. If Y denotes the number of defective tapes in the sample, then $P(Y \geq 1) = 1 - P(Y = 0)$ $= 1 - (1 - p)^n$, where p is the probability of observing a defective tape. This probability is to be 0.5. With $p = 0.1$, the equation becomes $1 - (0.9)^n = 0.5$, which has $n = 7$ as its closest solution.

b. With $p = 0.05$, the equation is $1 - (0.95)^n = 0.5$, which has $n = 14$ as its closest solution.

19. a. If the firm's gain is denoted by G and the number of wells that produce oil is denoted by Y, then $G = 1{,}000{,}000Y - 500{,}000$, since it costs \$50,000 to drill each of ten wells. Then, $E(G) = 1{,}000{,}000E(Y) - 500{,}000 = 1{,}000{,}000(10)(0.1) - 500{,}000 = \$500{,}000$.

b. $SD(G) = 1{,}000{,}000SD(Y)$
$= 1{,}000{,}000\sqrt{10(0.1)(0.9)}$
$= \$948{,}683$

c. Yes; there is a good chance the firm will lose money since 0 is only about one half a standard deviation below the expected gain. Yes; there is s good chance that the firm can make \$1.5 million, since this figure is only about 1 standard deviation above the expected gain.

> **iii.** What is the approximate probability of seeing at least 10 of the 16 households with income below \$39,000 annually? Show two different methods of answering this question.
>
> **c.** Suppose in a sample of 16 households none had an annual income below \$39,000. What might you suspect about this sample?
>
> **17.** A home alarm system has detectors covering n zones of the house. Suppose the probability is 0.7 that a detector sounds an alarm when an intruder passes through its zone, and this probability is the same for all detectors. The alarms operate independently. An intruder enters the house and passes through all the zones.
>
> **a.** What is the probability that an alarm sounds if $n = 3$?
>
> **b.** What is the probability that an alarm sounds if $n = 6$? Is the probability of part a doubled?
>
> **c.** Suppose the home owner wants the probability that an alarm will sound to be about 0.99. How large must n be?
>
> **18.** From a large lot of new video tapes, n are to be sampled by a potential buyer and the number of defective tapes Y is to be observed. If at least one defective tape is seen in the sample of n tapes, the potential buyer will reject the entire lot. Find n so that the probability of detecting at least one defective tape is 0.5 in each case.
>
> **a.** 10% of the lot is defective.
>
> **b.** 5% of the lot is defective.
>
> **19.** An oil-exploration firm is to drill 10 wells. Each well has a probability of 0.1 of producing oil. It costs the firm \$50,000 to drill each well. A successful well will bring in oil worth \$1,000,000.
>
> **a.** Find the firm's expected gain from the 10 wells.
>
> **b.** Find the standard deviation of the firm's gain for the 10 wells.
>
> **c.** Is there a good chance that the firm will lose money on the 10 wells? A rough approximation is acceptable here.
>
> **d.** Is there a good chance that the firm will gain more than \$1.5 million from the 10 wells?
>
> **20.** Ten CD players of a model no longer made are to be sold
>
> 74 LESSON 8

20. If Y denotes the number of CD players that fail in the first month, Y has a binomial distribution with $n = 10$ and $p = 0.08$.
The gain G for the seller is $G = 100(10) - 200Y$, since the seller sells all ten players for \$100 each. It follows that $E(G) = 100(10) - 200(10)(0.08) = 1000 - 160 = \$840$.

STUDENT PAGE 75

for \$100 each with a "double your money back" guarantee if the CD player fails in the first month of use. Suppose the probability of such a failure is 0.08. What is the expected net gain for the seller after all 10 CD players are sold? Ignore the original cost of the CD players to the seller.

SUMMARY

For a random sample of size n from a population with probability of "success" p on each selection, the probability that the number of "successes" in the sample Y equals a specific count c is given by

$$P(Y = c) = \binom{n}{c} p^c (1 - p)^{n-c}$$

where $\binom{n}{c}$ is the binomial coefficient given by

$$\binom{n}{c} = \frac{n!}{c!(n-c)}$$

The expected value of Y and the standard deviation of Y are given, respectively, by

$$E(Y) = np$$

and

$$SD(Y) = \sqrt{npq}$$

The random variable Y is said to have a ***binomial distribution***.

LESSON 9

The Geometric Distribution

Materials: 200 pennies for each group
Technology: graphing calculators or computers (optional)
Pacing: 1–2 class periods

Overview

Whereas the number of successes in a fixed number of independent trials leads to the binomial distribution, the sequential number on which the first success occurs leads to the geometric distribution. The development of the geometric distribution uses the same building blocks as the binomial, namely, a sequence of independent trials in which all trials have the same probability of success. The geometric is the simplest of the *waiting-time* distributions and has many applications in that area.

Several examples of waiting-time situations are counting the days until it rains or until the sun shines, counting the minutes until the telephone rings, counting the times you have to take an exam until you pass, and counting the number of extra points made by a place-kicker in football until he misses. Note in the latter example that success does not have to be a positive occurrence; it is simply the outcome you are looking for in the counting process.

Teaching Notes

This distribution has an infinite number of possible outcomes and, hence, derivation of its properties requires work with infinite series. Thus, it may be appropriate to review some basic ideas of infinite series, or introduce them if the students have not seen them before.

There are two simulations introduced in this lesson. The first is similar in spirit to the ones for the binomial distribution and can be done with either random digits or a physical device such as a spinner. The second requires sets of pennies and should not be programmed for the computer or calculator. Discuss with the students how these simulations differ and how they are similar. Both lead to the geometric distribution.

This lesson can be completed in one or two class periods if both simulations are done by all groups. An alternative is to assign one simulation to the groups as homework and then have students report the results in class.

Technology

Once the geometric distribution is developed, the computations are fairly easy and do not require much technology. The first simulation can be done with the graphing calculator or computer, but can also be easily accomplished by hand.

Follow-Up

There are many applications of the geometric distribution, such as the coupon-collecting problem in the assessment for Lessons 8 and 9. Have students find similar contests or games and ask them to use the geometric distribution to model how long it might take them to win.

STUDENT PAGE 76

LESSON 9

The Geometric Distribution

What is the probability that your team will have to play four games before it gets its first loss?

What is the general shape of waiting-time distributions?

How can a mathematical model be developed for waiting-time distributions?

OBJECTIVES

Understand the basic properties of the geometric distribution.

Use the geometric distribution as a model for certain types of counts.

The opening example of Lesson 8 considered the problem of how many high-school graduates should appear in a random sample of four adults from a city. Suppose, however, that the interviewer is interested in finding only one high-school graduate from this city. A question of interest in this case might be, "How many people must be selected until a high-school graduate is found?" In other words, "How long will we have to wait to find the first success?" The first person selected might be a high-school graduate but, if not, a second person will have to be interviewed, and so on.

INVESTIGATE

How Long Must We Wait?

Describe another situation in which waiting time until "success" is an important consideration.

Discussion and Practice

1. Recall that the proportion of high-school graduates in the population is 0.80.

STUDENT PAGE 77

a. Can the number of adults that must be selected until the first high-school graduate is found be determined in advance of the sampling?

b. Could the number of adults selected until the first high-school graduate is found be quite large?

c. Is there a high probability that the number of adults selected until the first high-school graduate is found will be quite large? Explain your reasoning.

We will begin our investigation of the problem of waiting for a success by constructing a simulation for the situation presented above and then moving on to other cases. The simulations can be completed most efficiently if you work in pairs.

2. If you select adults one at a time, how many must be selected until the first high-school graduate is found? Recall that the proportion of high-school graduates in the population is 0.80. The number of the trial on which the first success occurs is a random variable, which we will denote by Y. Its distribution can be approximated by the following simulation.

a. What is the probability that a high-school graduate is seen on the first selection? Find a device that will generate an event with this probability. You may use random-number tables, a random-number generator in your calculator, or some other device.

b. Simulate the selection of the first adult by generating an event with the device selected in part a. Did you see a high-school graduate? In other words, was your waiting time to success just one interview?

c. If you had success on the first selection, then stop this run of the simulation. If you did not have success on the first trial, then continue generating events until the first success occurs. Record a value for Y, the number of the trial on which the first success occurred. NOTE: This number must be at least 1.

d. Repeat parts b and c ten times, recording a value for Y each time.

e. Combine your values of Y with those of your classmates.

THE GEOMETRIC DISTRIBUTION 77

Solution Key

Discussion and Practice

1. **a.** No; theoretically, the sampling could go on for a long time before the first high-school graduate is found, even though there is an 80% chance of finding a high-school graduate on any one selection.

b. Yes; in light of the answer to Problem 1a, the sample size required to find the first high-school graduate could be quite large.

c. No; since the chance of finding a high-school graduate on the first selection is 0.8, the chance of finding at least one high-school graduate among the first few selections should be quite large.

2. **a–e.** Various devices can be used for generating this simulated data, including random numbers or colored chips (8 red and 2 white). The data displayed in this answer were produced with the random-number generator in a computer. In a total of 99 trials, success was found on the first trial 81 times, on the second trial 15 times, and on the third trial 3 times. In this simulation, it never took more than 3 trials to find the first successful outcome. The estimated probabilities for the three observed outcomes are simply the relative frequencies with which they occurred.

Y	Frequency	$P(Y)$
1	81	0.818182
2	15	0.151515
3	3	0.030303

The distribution of the values of Y are pictured in the graph below. The distribution is highly skewed toward the higher values, as it should be for a distribution in which a small value is quite likely to occur but a large value has only a small chance of occurring.

The mean of this distribution is a little greater than 1, being pulled in that direction by the long tail toward the larger values. Actually, the mean for this distribution is 1.21.

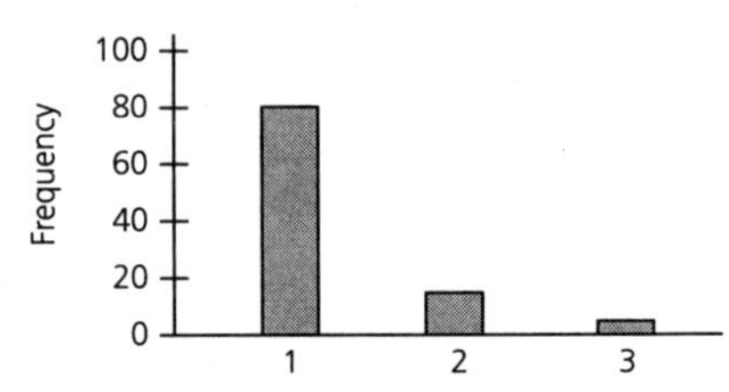

3. a–d. This activity should be done with real pennies, not with a computer or calculator simulation. The computer or calculator can be used to keep track of the data.

The data below are the result of one such simulation with 200 pennies.

Y	Frequency	$P(Y)$
1	102	0.510
2	48	0.240
3	28	0.140
4	11	0.055
5	3	0.015

Of the 200 coins, 102 turned up heads on the first toss. Of the 98 tails, 48 turned up heads on the second toss, and so on. Again, the probabilities for the values of Y are estimated by the relative frequency of outcomes out of the 200 coins. Thus, it is estimated from the data that the probability that a "fly" will die in the first day is 0.51, quite close to the theoretical value of 0.50 and the probability of dying in the second day is 0.24. Note that there are some "flies" that lived longer than five days.

The graph of the simulated probability distribution for Y, the age of the flies in days, is shown below. This distribution is highly skewed toward more days, which is to be expected. The mean will be pulled upward from 1 quite strongly by the long tail to the right. In this case, the mean, or estimated expected value of Y, is 1.79. Remember, however, that there were 8 "flies" still living when the simulation was stopped, and these would increase the mean even more.

STUDENT PAGE 78

i. Construct a plot to represent the simulated distribution of Y.

ii. Describe the shape of this distribution.

iii. Approximate the mean of this distribution.

3. Each fly of a certain species has a 0.5 chance of dying in any one-day period after its birth. Simulate the probability distribution of Y, the age of flies of this species. In this simulation, "age" can be thought of as the number of days until death, or the waiting time until death. Instead of generating the values of Y one at a time, as we did in Problem 2, consider a simulation that gives a whole set of Y values simultaneously.

a. Gather a set of about 200 pennies. Place the pennies in a container, shake well, and toss them onto a table. Count the heads.

b. If "head on a coin" represents "death of a fly" what fraction of flies died during the first day? Does this seem like a reasonable estimate of the probability that a fly died during the first day, $Y = 1$? Explain.

c. Gather the pennies that came up tails on the first toss. Place them in the container, shake well, and toss them onto a table. Count the heads. What fraction of the *original number* of pennies showed the first head on the second toss? What is the approximate probability that a fly dies during the second day, $Y = 2$?

d. Continue gathering the tails, tossing them, and counting the heads until there are five or fewer tails left.

i. Record the values of Y and the approximate probability for each.

ii. Show the simulated probability distribution for Y on an appropriate graph.

iii. Describe the shape of this distribution.

iv. Approximate the mean of this distribution.

4. Compare the methods of simulation used in Problems 2 and 3. What would be the difficulty of using the method of Problem 3 on Problem 2?

5. To provide more experience with probability distributions of waiting times, we have simulated the cases for $p = 0.3$, $p = 0.5$, and $p = 0.8$, where p is the probability of success

78 LESSON 9

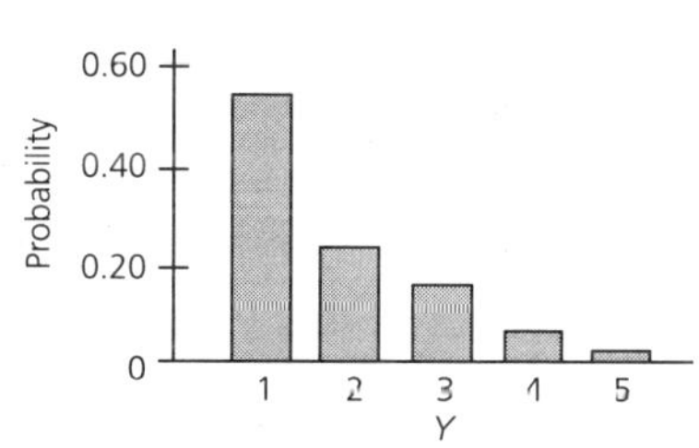

4. The two methods of simulation are quite similar and either could be used on the first problem. The event "fly dies when two days old" is essentially the same type of event as "it takes two selections to find the first high-school graduate." The problem in using the second method (coins) on the first problem is that you would have to find something that can be selected out with probability 0.8 at each stage. Ten-sided dice will work, for example.

5. **a.** The shapes are similar in that all three are highly skewed toward the greater values. The shapes differ in that there is a greater probability at 1 and a more pronounced skewness as p increases.

b. Although difficult to see on the graphs, the probabilities do not quite add to 1, since there is always a small chance that the number of trials until the first success could be greater than any value yet observed.

c. The means are approximately 3.0, 2.0, and 1.2, respectively.

STUDENT PAGE 79

on any one event. These are shown below. The latter two cases should have similarities to those generated in Problems 2 and 3.

a. How are the shapes of the three distributions similar? How are they different?

b. Do the probabilities shown in each distribution appear to add to 1? If not, why not?

c. Approximate the mean of each distribution.

Simulated Waiting-Timing Distributions

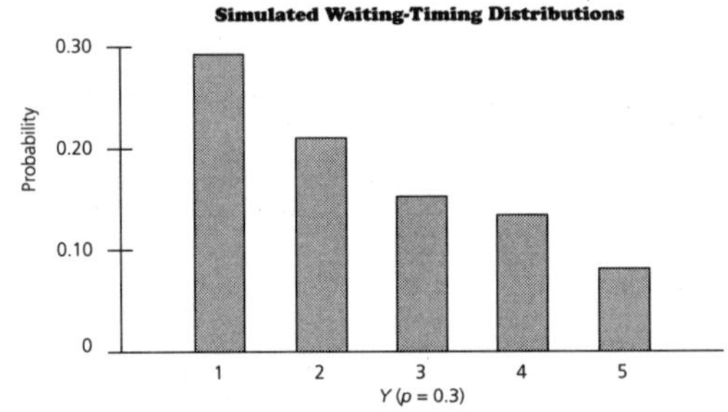

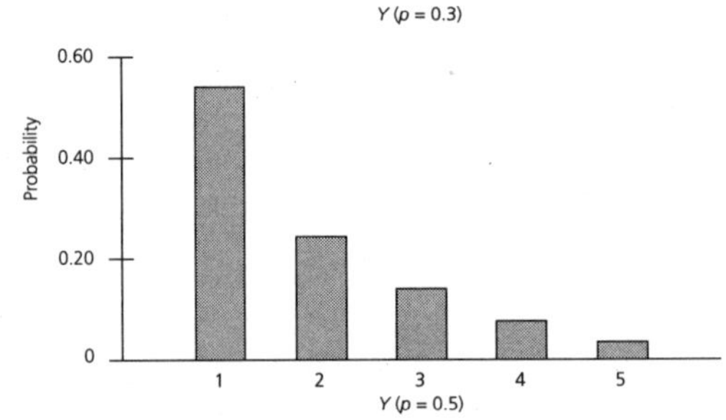

THE GEOMETRIC DISTRIBUTION **79**

6. a. $P(Y = 2)$
$= P[(X_1 = 0) \text{ and } (X_2 = 1)]$
$= P(X_1 = 0) \cdot P(X_2 = 1)$
$= qp$
$P(Y = 3)$
$= P[(X_1 = 0) \text{ and } (X_2 = 0) \text{ and } (X_3 = 1)]$
$= P(X_1 = 0) \cdot P(X_2 = 0) \cdot P(X_3 = 1)$
$= q^2p$

b and c. $P(Y = c) = P(X_1 = 0) \cdot P(X_2 = 0) \cdot \ldots \cdot P(X_{c-1} = 0) \cdot P(X_c = 1) = q^{c-1}p$ for any positive integer c. For Y to take on the value c, the first $c - 1$ trials must be failures followed by the first success on trial numbered c.

STUDENT PAGE 80

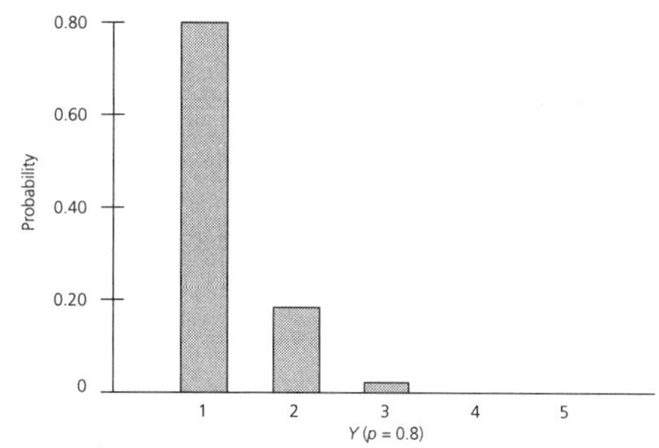

The Geometric Distribution

Simulation provides a good way to investigate properties of distributions, but it is inefficient to run a simulation every time a new problem involving one of these waiting-time distributions arises. It turns out that a formula for the probability distribution of Y can be derived quite easily—more easily, in fact, than in the binomial case of Lesson 8.

Suppose we are sampling people from a large population in which the proportion of people having the characteristic labeled success is p. Using notation similar to that of Lesson 8, we can record the outcome of each selection in sequence by letting

$X_i = 1$ if the ith selection in the sequence is a success

and

$X_i = 0$ if the ith selection in the sequence is not a success.

Then, $P(X_i = 1) = p$ and $P(X_i = 0) = 1 - p = q$.

6. Development of a formula for the probability distribution of Y, the selection on which the first success occurs, proceeds along the following lines.

a. Write $P(Y = 2)$ in terms of the random variables X_i. Do the same for $P(Y = 3)$.

b. Write $P(Y = c)$ in terms of the random variables X_i, for any positive integer c.

80 LESSON 9

(6) d. The probabilities of the successes or failures on the individual trials are multiplied together. This is allowed only if the trials are independent of one another. Independence would not be the case, for example, if names were to be selected without replacement among the students in the classroom until the first female is selected. "Without replacement" means that each selected name is kept out of the pool while the next one is drawn.

7. a. Following the notation used on the probability distribution of Y, $E(Y) = 1(p) + 2(qp) + 3(q^2p) + 4(q^3p) + \ldots$ Factoring out a p and writing the terms in a triangular array gives the answer.

STUDENT PAGE 81

c. Using the result of part b, write $P(Y = c)$ in terms of p and q. Make the result as compact as possible.

d. What assumption is necessary for the probability calculations used in part c?

With the answer to Problem 6, you can now see what the probability distribution for the random variable Y looks like in table form.

Y	P(Y)
1	p
2	qp
3	q^2p
4	q^3p
.	.
.	.
.	.

The sum of these probabilities form an infinite series, and the sum of the probabilities across all possible values of Y form an infinte sum:

$$\sum_{c=1}^{\infty} q^{c-1}p = p\sum_{c=1}^{\infty} q^{c-1} = p(1 + q + q^2 + q^3 + q^4 + \ldots) = p\left(\frac{1}{1-q}\right) = 1.$$

The infinite sum inside the brackets is called a ***geometric progression***. The basic result for summing such a series is as follows:

$$a + ax + ax^2 + ax^3 + \ldots = \frac{a}{1-x}$$

as long as the absolute value of x is less than 1.

You see, now, why the probability distribution developed in this lesson is called the ***geometric distribution***.

7. The expected value of the geometric distribution can be found by summing an array of infinite series. Show that $E(Y)$, for the geometric random variable Y, can be written in a triangular array as

$$\begin{aligned} E(Y) = p(1 &+ q + q^2 + q^3 + \ldots \\ &+ q + q^2 + q^3 + \ldots \\ &+ q^2 + q^3 + \ldots \\ &+ q^3 + \ldots \\ &+ \ldots) \end{aligned}$$

a. Sum each row of the array inside the brackets by using the result for geometric progressions.

THE GEOMETRIC DISTRIBUTION **81**

b. The sums for the first four rows in the triangular array, as shown in the diagram, are $\frac{1}{p}, \frac{q}{p}, \frac{q^2}{p}, \frac{q^3}{p}$ and with similar terms to follow in the infinite array.

c. $E(Y) = p(\frac{1}{p} + \frac{q}{p} + \frac{q^2}{p} + \frac{q^3}{p} \ldots)$

$= 1 + q + q^2 + q^3 + \ldots$

$= \frac{1}{1-q} = \frac{1}{p}$

d. The expected number of trials until the first success is the reciprocal of the probability of success on any one trial. Thus, if you toss a balanced coin, you would expect to take two tosses to see the first head. If you make only 40% of your free throws in basketball, you would expect to wait an average of 2.5 throws until your first successful attempt, if the throws are independent.

8. a–c.

p	0.3	0.5	0.8
$E(Y)$	3.33	2.00	1.25
$SD(Y)$	2.79	1.41	0.56
Interval	(0.54, 6.12)	(0.59, 3.41)	(0.69, 1.81)

The table shows the values of p for the three graphs of Figure 1 and then provides the expected values, standard deviations, and the interval of one standard deviation to either side of the expected value. Note that for small values of p the standard deviation and the interval are quite large and do not provide much guidance as to where most of the values of Y would lie. The standard deviation is a useful guide to where the values of Y are concentrated only for situations involving a large value of p.

STUDENT PAGE 82

b. Find $E(Y)$ by summing the row totals you just found, again using the result for geometric progressions.

c. Does your result for $E(Y)$ make sense from a practical point of view? Explain.

8. The standard deviation of a geometric random variable Y is difficult to find from the basic distribution, but it turns out to be $SD(Y) = \frac{1}{p}\sqrt{1-p}$.

a. Find the expected value and standard deviation for each of the three geometric distributions simulated on pages 79–80.

b. Plot the expected value and points one standard deviation above and below the expected value on each distribution on pages 79–80.

c. Does the standard deviation appear to be a good measure of a typical deviation from the mean here? Explain.

Practice and Applications

9. In the scenario outlined at the beginning of this lesson, the probability of randomly selecting a high-school graduate was 0.8.

a. What is the probability of getting the first high-school graduate on the third selection?

b. What is the probability that it will take at least three selections to get the first high-school graduate?

c. Suppose the first high-school graduate comes up on the third selection. What is the probability that it will take at least three more selections to obtain a second high-school graduate?

d. What is the expected number of selections to obtain the first high-school graduate?

e. What is the expected number of selections to obtain two high-school graduates?

10. Suppose 10% of the engines manufactured on a certain assembly line have at least one defect. Engines are randomly sampled from this line one at a time and tested.

a. What is the probability that the first non-defective engine is found on the third trial?

b. What is the expected number of engines that need to be

82 LESSON 9

Practice and Applications

9. a. $P(Y = 3) = (0.2)^2 (0.8) = 0.032$

b. $P(Y \geq 3) = 1 - P(Y = 1) - P(Y = 2) = 1 - 0.8 - 0.2(0.8) - 0.04$

c. The answer is the same as for Problem 9b since the process of waiting starts over once the first success is found. The trials remain independent with a constant probability of 0.8 for success.

d. $E(Y) = \frac{1}{p} = \frac{1}{0.8} = 1.25$

e. The expected wait for two graduates is twice the wait for one, because of the reasoning of Problem 9c. Thus, the answer is 1.25 + 1.25 – 2.50.

10. a. You are looking for non-defective engines. Therefore, $p = 0.9$, the probability of selecting a non-defective engine.
$P(Y = 3) = (0.1)(0.1)(0.9) = 0.009$

b. $E(Y) = \frac{1}{0.9} = 1.11$

(10) **c.** $SD(Y) = \frac{\sqrt{0.1}}{0.9} = 0.35$

d. Letting cost be denoted by C, $C = 100Y$, since it costs $100 to test each engine. Thus, $E(C) = 100E(Y) = 100(1.11)$ $= \$111$ and $SD(C) = 100SD(Y)$ $= 100(0.35) = \$35$.

The figure of $200 is about 2.5 standard deviations above the mean, and a point that far away from the mean will not be reached often.

11. a. If the lines are busy 60% of the time, the chance of getting a call through is only $p = 0.4$. Thus, $P(Y) = 1) = 0.4$, $P(Y) = 2) = 0.6(0.4) = 0.24$, and $P(Y) = 4) = (0.6)^3(0.4) = 0.0864$.

b. $E(Y) = \frac{1}{0.4} = 2.5$

12. a. You are looking for a successful well, and $p = 0.1$. Thus, $E(Y) = \frac{1}{0.1} = 10$.

b. Since each well costs $50,000 to drill, the cost of drilling C is given by $C = 50{,}000Y$. Thus, $E(C) = 50{,}000E(Y)$ $= 50{,}000(10)$ $= \$500{,}000$ and $SD(C) = 50{,}000SD(Y)$ $= 50{,}000\left(\frac{\sqrt{0.9}}{0.1}\right)$ $\approx \$474{,}342$.

Notice that the SD is almost as large as the expected cost. There is great variability in this process.

c. The probabilities for these "at least" statements can be worked out directly, but it is useful to develop a general formula for such problems. In general,

$P(Y$ is at least $c)$
$= q^{c-1}p + q^cp + q^{c+1}p + \ldots$
$= p(q^{c-1} + q^c + q^{c+1} + \ldots)$
$= p\left(\frac{q^{c-1}}{p}\right) = q^{c-1}$.

The condition that $Y \geq c$ means that there must have been $c - 1$ failures up to this point. For the problem at hand, $P(Y \geq 5) = (0.9)^4 = 0.6561$ and $P(Y \geq 15) = (0.9)^{14} = 0.2288$.

STUDENT PAGE 83

tested before the first non-defective engine is found?

c. What is the standard deviation of the number of engines that need to be tested before the first non-defective engine is found?

d. Suppose it costs $100 to test one engine. What are the expected value and the standard deviation of the cost of inspection up to and including the first non-defective engine? Will the cost of inspection often exceed $200? Explain.

11. The telephone lines coming into an airline reservation office are all busy about 60% of the time.

a. If you are calling this office, what is the probability that it will take you only one try to get through? Two tries? Four tries?

b. What is your expected number of tries to complete the call?

12. An oil-exploration firm is to drill wells at a particular site until it finds one that will produce oil. Each well has a probability of 0.1 of producing oil. It costs the firm $50,000 to drill each well.

a. What is the expected number of wells to be drilled?

b. What are the expected value and the standard deviation of the cost of drilling to get the first successful well?

c. What is the probability that it will take at least five tries to get the first successful well? At least 15?

SUMMARY

For a sequence of selections from a large population in which the probability of "success" p stays the same for all selections, the number of the selection on which the first success occurs Y has a ***geometric distribution***. The mean and the standard deviation of this distribution are given by

$$E(Y) = \frac{1}{p}$$

and

$$SD(Y) = \frac{1}{p}\sqrt{1-p}$$

THE GEOMETRIC DISTRIBUTION **83**

ASSESSMENT

Lessons 8 and 9

STUDENT PAGE 84

ASSESSMENT

Lessons 8 and 9

1. In the population as a whole, about 46% of people have type-O blood, about 42% have type-A, about 8% have type-B, and about 4% have type-AB.
   a. In a random sample of five people, how many would be expected to have type-A blood? How many would be expected to have either type-A or type-AB blood?
   b. For a random sample of five people, find the probability distribution for the number having type-AB blood. Describe the shape of this distribution.
   c. A total of 200 people are to donate blood at a certain blood bank this week. Find the expected value and the standard deviation of the number of type-A donors the blood bank will see this week.
   d. The blood bank in part c needs 100 donors of type-A blood this week. If 200 donors appear this week, is there a good chance that it will get the number it needs? Explain your reasoning.
   e. If the donors to a blood bank come in sequential order, what is the chance that more than four donors will have to be tested before the first one with type-A blood shows up? Answer the same question for type-AB blood.
   f. If the donors to a blood bank come in sequential order, what is the expected number of donors that must be tested in order to find the first one with type-AB blood?
2. Refer to the percents of blood types given in Problem 1. The Rh, or *Rhesus*, factor in the blood is independent of the blood type. About 85% of people are Rh positive.
   a. For a random sample of five donors, find the probability distribution of the number of type O-negative donors that will be seen. Describe the shape of this distribution.
   b. A total of 200 people are to donate blood in a certain blood bank this week. Find the expected value and the standard deviation of the number of A-positive donors.

84 ASSESSMENT

Solution Key

1. **a.** Let Y denote the number of donors with type-A blood. For this case, $p = 0.42$ and $E(Y) = 5(0.42) = 2.1$.

   The probability that a random donor has type-A or type-AB blood is $0.42 + 0.04 = 0.46$. The expected number of donors, out of the five, with type-A or type-AB blood is $5(0.46) = 2.3$.

   b. For $n = 5$ and $p = 0.42$, the distribution of the number of successes is shown below:

Y	$P(Y)$
1	0.06564
2	0.23765
3	0.34418
4	0.24923
5	0.01307

   This probability distribution is fairly symmetric, but is slightly skewed toward the greater values, since p is slightly below 0.5.

   c. Let Y denote the number of type-A donors in a sample of 200 with $p = 0.42$. Then, $E(Y) = 200(0.42) = 84$ and $SD(Y) = \sqrt{200(0.42)(0.58)} \approx 7.0$.

   d. The 100-donor goal is 2.3 standard deviations above the mean under a distribution which would look nearly normal. A value this far away from the mean has a very small chance of occurring.

   e. P(number of trials until first success is greater than c) $= (1 - p)^c$.

   For the case of type-A blood, $p = 0.42$, and the probability of waiting through more than four donors for the first success is $(0.58)^4 = 0.1132$.

   For the case of type-AB blood, $p = 0.04$, and the probability of waiting through more than four donors for the first success is $(0.96)^4 = 0.8493$.

   f. E(waiting period for a type-AB donor) $= \frac{1}{0.04} = 25$. The blood bank could wait quite a long time to find a donor with a rare blood type.

2. **a.** For counting Y, the number of O-donors, the probability of success is $p = (0.46)(0.15) = 0.069$ because of the independence of the two conditions. For $n = 5$, the distribution of Y is given at right:

Y	$P(Y)$
1	0.699437
2	0.038419
3	0.002847
4	0.000106
5	0.000002

   This probability distribution is extremely skewed toward the greater values because of the very small chance of success on any one randomly selected donor.

STUDENT PAGE 85

If the blood bank needs 80 A-positive donors this week, does it stand a good chance of getting them? Explain.

c. The blood bank is in need of an A-positive donor as soon as possible. Describe the distribution of the number of donors that must be tested to find the first A-positive donor. What is the expected number of donors to be tested in order to find one who is A positive?

3. You are to take a ten-question multiple choice exam. Each question has four choices, of which only one is correct. You know none of the answers and decide to guess at an answer for each question.

a. Describe how to set up a simulation for the probability distribution of the number of questions answered correctly.

b. What is the probability that you will answer at least 60% of the questions correctly?

c. Suppose the teacher changes the questions so that each one has three choices, one of which is correct. Would your chance of answering at least 60% correctly by guessing go up or down from the answer on Problem 3b? Explain.

d. Refer to the original scenario of four choices per question. Suppose the test now has five questions rather than ten. Does your chance of getting at least 60% correct by guessing increase or decrease from your chance in Problem 3b? Explain.

4. Each box of a certain brand of cereal contains a coupon that can be redeemed for a poster of a famous sports figure. There are five different coupons, representing five different sports figures.

a. Suppose you are interested in one sports figure in particular. Set up a simulation that would produce an approximate probability distribution for the number of boxes of cereal you would have to buy in order to get one coupon for that particular poster. You stop buying cereal when you get the poster you want.

b. What is the probability that you would get the coupon for the particular sports figure of interest in four or fewer boxes of cereal? You stop buying cereal when you get the poster you want.

LESSONS 8 AND 9 85

b. For Y equal to the number of A-positive donors in a sample of 200 with $p = (0.42)(0.85) = 0.357$, $E(Y) = 200(0.357) = 71.4$, and $SD(Y) = \sqrt{200(0.357)(0.643)} \approx 6.8$.

In this case, the target of 80 A-positive donors is only about 1.3 standard deviations above the mean under a distribution that is nearly normal. A value this far from the mean is quite likely to occur, and the blood bank has a good chance of achieving the target with 200 donors.

c. If T denotes the number of tests that must be run until the first A-positive donor is found, then T has a geometric distribution with $p = 0.357$. In symbols, $P(T = c) = (0.643)^c(0.357)$ for any positive integer c and $E(T) = \frac{1}{0.357} = 2.80$.

3. **a.** Possible answer: The probability of a correct choice is $p = \frac{1}{4} = 0.25$. Random digits can be used to simulate an event with this probability. One way to do it is to disregard two of the digits, say the 9 and 0, and use 1 and 2 to indicate success among digits sampled from 1 through 8. Another randomization model is to toss two coins and let success be denoted by the occurrence of two heads. After a randomization device is selected, it must be employed ten times to simulate one exam. The number of successes is counted for this simulated exam. Then, the process is repeated many times to simulate the distribution of the number of successes.

b. With $n = 10$, $p = 0.25$, and Y defined as the number of successes, the distribution of Y is as given at right:

Y	$P(Y)$
0	0.056314
1	0.187712
2	0.281568
3	0.250282
4	0.145998
5	0.058399
6	0.016222
7	0.003090
8	0.000386
9	0.000029
10	0.000001

The chance of answering at least 60%, or 6 questions, correctly is the sum of the last five rows of the table, which is 0.0197. The student has a very slight chance of guessing his or her way to a passing grade.

c. Now, the chance of a correct answer on any one question is $\frac{1}{3}$. The probability distribution for the number of correct answers out of ten questions is given by:

Y	$P(Y)$
0	0.018228
1	0.089782
2	0.198993
3	0.261365
4	0.225281
5	0.133151
6	0.054652
7	0.015382
8	0.002841
9	0.000311
10	0.000015

From this distribution, the probability of getting 6 or more answers correct is 0.0732, a little larger than in the case of four choices, but not much. The student still has a poor chance of guessing to a passing grade.

d. The distribution of a number of correct answers Y for $n = 5$ and $p = 0.25$ is shown below.

Y	$P(Y)$
0	0.237305
1	0.395508
2	0.263672
3	0.087891
4	0.014648
5	0.000977

Now, getting at least 60% correct is the same as getting 3 or more correct answers, which has probability 0.1035. The chance of obtaining a passing grade by guessing is much greater on the short exam than on the long one.

4. **a.** You are interested in one particular coupon out of five. Thus, your probability of success on any one purchase is $\frac{1}{5} = 0.2$, assuming the coupons are randomly distributed in equal numbers. You could sample random digits, counting the number of digits until the first 0 or 1 appears. This process should be repeated many times in order to generate a distribution of values for the number of coupons necessary to achieve success.

b. P(purchase 4 or fewer boxes) $= 1 - P$(the coupon of interest is not in any of the first four boxes) $= 1 - (0.8)^4 = 1 - 0.4096 = 0.5904$.

c. E(number of boxes purchased)

$$= \frac{10}{0.2} = 5.$$

d. In the first stage, either of two coupons will denote success, and so the probability of success is $\frac{2}{5} = 0.4$. After one of these is obtained, the process starts over and the other coupon must be obtained, which means the probability of success is 0.2.

For the two stages together, E(number of boxes purchased) $= E$(number of boxes purchased on 1st stage) $+ E$(number of boxes purchased on 2nd stage)

$$= \frac{1}{0.4} + \frac{10}{0.2} = 2.5 + 5 = 7.5.$$

STUDENT PAGE 86

c. What is the expected number of boxes of cereal you would have to buy to get the one coupon you want?

d. Suppose you want to get two particular posters out of the five available. What is the expected number of boxes of cereal you would have to buy to get the two specific ones? (HINT: At the outset, the probability of getting a coupon you want in any one box is $\frac{2}{5}$. After you get one of them, what is the probability of getting the other coupon you want? Think of the expected number of boxes of cereal being purchased in terms of these two stages.)

86 ASSESSMENT

Teacher Resources

NAME ______________________________

1. A quality-improvement plan for an aircraft-repair facility required information on the number of defects seen on repaired pumps coming from the hydraulics shop. A random sample of 20 pumps from the shop showed the following counts on the number of defects per pump:

6 3 4 0 2 7 3 1 0 0 4 3 2 2 6 5 0 7 2 1

The random variable of interest for planning future inspections is Y, the number of defects per pump.

a. Use the data from the sample of 20 pumps to approximate a probability distribution for the random variable Y.

b. Find the expected value of Y by using the approximate distribution for Y found in part a.

c. Find the standard deviation of Y by using the approximate distribution found in part a.

2. Each defect found in a pump as described in Problem 1—usually a faulty gasket—costs $50 to repair.

a. Find the estimated expected repair cost per pump.

b. Find the estimated standard deviation of repair costs per pump.

c. Suppose $200 per pump is budgeted for repairing these defects. Will this amount be exceeded often? Explain your reasoning.

3. The table below shows the frequency, in millions, for ages of cars in use in the United States for 1980 and 1994. For example, in 1980 there were 52.3 million cars under 5 years old in use across the country.

Age in years	1980	1994
5 or less	52.3	45.4
6 - 8	25.2	27.7
9 -11	14.6	25.1
12 or more	12.5	31.4

Source: *Statistical Abstract of the United States,* 1996

a. Construct plots of each of these distributions. How do the two distributions differ?

b. Approximate the mean of each distribution.

c. If you could obtain more detailed information on the ages of cars in use, do you think the means would increase, compared to your approximations? Explain your reasoning.

d. Approximate the medians for each distribution. How do they compare?

e. Explain which distribution must have the larger standard deviation without doing the calculations.

4. A merchant stocks a certain perishable item. On any given day, she will have a demand for either two, three, or four of these items with probabilities 0.1, 0.4, and 0.5, respectively. She buys each item for $1.00 and sells each item for $1.20. Any items left over at the end of the day represent a total loss. How many items should the merchant stock in order to maximize her expected daily profit?

NAME ______________________________

1. A quality-improvement plan in an aircraft repair facility calls for sampling a number of hydraulic pumps that have been rebuilt and counting the number of defects that are still present. These defects are usually the result of a faulty gasket. The following table shows the approximate probability distribution of Y, the number of defects observed per pump.

Y	$P(Y)$
0	0.20
1	0.10
2	0.20
3	0.15
4	0.10
5	0.05
6	0.10
7	0.10

a. The next group of aircraft coming in to be repaired will have a total of 100 pumps that need rebuilding. Find the expected value and the standard deviation of the mean number of defects per rebuilt pump that can be anticipated in this sample of 100 pumps.

b. The quality improvement plan set a goal of no more than 230 defects to be observed on the 100 rebuilt pumps. Would the facility be likely to reach that goal if it is still operating under the distribution of defects per pump shown above? Explain your reasoning.

2. The table below shows the frequency distribution of the ages of drivers under the age of 25, called "young adults."

Age	Frequency
16	1,470,521
17	2,200,842
18	2,493,137
19	2,727,972
20	2,836,091
21	2,921,521
22	3,130,025
23	3,482,642
24	3,608,980

Source: *The World Almanac,* 1997

a. If Y denotes the age of a driver randomly selected from the population of young-adult drivers, construct a probability distribution for Y.

b. A polling company is to randomly sample 400 young-adult drivers. Describe the distribution of the possible values of the mean age of the drivers in this sample.

c. Suppose a sample of 400 young-adult drivers produced a mean age of 21 years. What questions would this raise in your mind and why?

3. In a recent poll of 1000 Americans, it was found that 81% expressed belief in the existence of heaven (*Time,* March 24, 1997).

a. What is the approximate margin of error for this poll result?

b. Explain what is meant by the "margin of error."

4. According to the U.S. Bureau of the Census, about 15% of Americans have no health insurance. A government agency is designing a sampling plan to interview Americans about their concerns. The plan calls for a sample size of 500.

a. If 500 Americans are randomly sampled, describe the probability distribution of the potential values of the sample proportion of those having no health insurance.

b. The agency would like to locate 100 Americans with no health insurance for further study. Is the sample of 500 likely to produce 100 with no health insurance? Explain your reasoning.

NAME ______________________________

1. According to the U.S. Bureau of the Census, 15% of Americans are without health insurance.

   a. Six new patients come through the emergency room of a hospital in one morning. Find the probability distribution of the number of new patients without health insurance. Make a rough plot of the distribution and comment on its shape. What assumption is necessary for your answer to be valid?

   b. Fifty new patients come through the emergency room of a hospital in one day. Find the expected value and standard deviation of the number of patients without health insurance among the 50. What assumption is necessary for your answer to be valid?

   c. Among the 50 new patients coming through the emergency room, what is the approximate probability that more than ten of them will be without health insurance? A rough approximation will be sufficient.

2. A quality-control plan in a plant manufacturing automobile batteries calls for the random selection and weighing of one battery every hour. Weight is an important quality measurement, because the quality of a battery depends on the amount of lead in its plates. According to the manufacturing specifications, these batteries should have approximately normally distributed weights with a mean of 68 pounds and a standard deviation of 0.5 pound.

   a. If the weight of a sampled battery lies more than 3 standard deviations from the specified mean, an "out-of-control" signal is sent to the engineer in charge and a series of further checks on the production process are begun. Find the probability that no out-of-control signals are sent in one 8-hour shift if the process is operating according to specifications.

b. Repeat part a with the change that out-of-control limits are now only 2 standard deviations from the specified mean.

3. Review the quality-control scenario of Problem 2 and assume the out-of-control limits are set at 3 standard deviations.

a. How many hours can the plant manager expect to wait until the first out-of-control signal appears, assuming that the plant is operating according to specifications?

b. Find the standard deviation of the number of batteries that must be checked until the first out-of-control signal sounds.

c. Is there a good chance that the plant may see no out-of-control signals in 600 hours of operation? Explain your reasoning.

4. Your local theater is distributing prizes in the large tubs of popcorn it sells during the reprise of the *Star Wars*® movies. The prizes are small models of *Star Wars* spacecraft. Three different models are randomly distributed in equal numbers, one model per tub of popcorn. How many tubs of popcorn can you expect to buy in order to obtain a full set of models, that is, at least one of each of the three types available? Explain how you arrived at this answer.

Note: *Star Wars* is a registered trademark of Lucasfilm, *LTD*.

1. **a.** The data on individual pump defect counts must be placed into a relative-frequency table, since probabilities are estimated by relative frequencies. The table could look like this one, although the frequencies do not need to be recorded.

Y	Frequency	$P(Y)$
0	4	0.20
1	2	0.10
2	4	0.20
3	3	0.15
4	2	0.10
5	1	0.05
6	2	0.10
7	2	0.10

b. $E(Y) \approx 2.90$

c. $SD(Y) \approx 2.28$

2. **a.** Repair cost per pump, denoted by C, is given by $C = 50Y$.

$E(C) = 50E(Y) = 50(2.9) = \$145$

b. The standard deviation of repair cost per pump is

$SD(C) = 50SD(Y) = 50(2.28) = \$114$

c. The $200 budgeted amount is about one half a standard deviation above the mean. This amount could be exceeded often.

3. **a.** Shown below are the relative-frequency, or proportion, plots for the data on ages of cars.

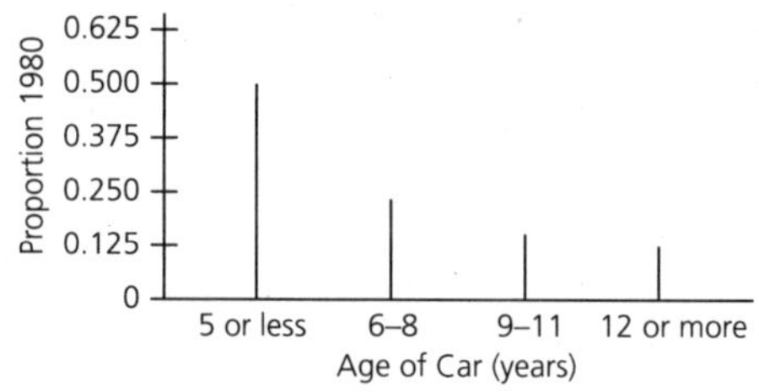

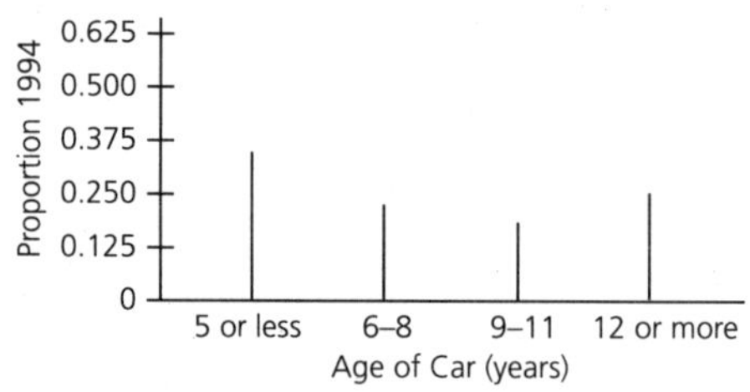

The distribution for 1980 is highly skewed toward the older cars, while the distribution for 1994 is much flatter and, in fact, increases slightly as we move toward the oldest age category. This shows that the cars in use in 1994 tend to be older than those in use in 1980, relatively speaking.

b. Using the midpoints of the intervals, 3, 7, and 10, as typical values and 14 as a typical value for the "12 or more" category, the means are 6.25 years and 7.88 years, respectively.

c. More information would probably show that there are cars in use that are much older than 12 years. In other words, the 14 used above might be too small. The means should increase with more information on the older cars becoming available.

d. The median for the 1980 ages is about 5 and the median for the 1994 ages is a little less than 7. The cars in use in 1994 appear to be older than the ones in use in 1980, on the whole.

The easiest way to see the locations of the medians is to turn the frequency data into relative-frequency data, or estimated probabilities. The result is shown below:

Age	$P(Y)$ '80	$P(Y)$ '94
5 or less	0.5000	0.3503
6–8	0.2409	0.2137
9–11	0.1396	0.1937
12 or more	0.1195	0.2423

e. The 1994 data have the more even frequencies in the categories and, hence, the greater spread. The 1980 data has a high percent of values concentrated in the 5-or-under category and relatively low percents in the two oldest categories.

4. The choice is to be made among two, three or four items as the number to stock. The merchant makes a profit of $0.20 on each item sold, but loses $1.00 if an item is not sold. Let G denote the merchant's gain. Consider the three choices separately.

Stock two: In this case, the two are sure to sell because there is a probability of 1 that at least two customers will show up. G has the distribution

G	$P(G)$
0.40	1

From this distribution, $E(G) = \$0.40$.

Stock three: There is a demand for at least three items with probability 0.9. If she stocks three and sells only two, she loses $0.60. G has the distribution

G	$P(G)$
−0.60	0.1
0.60	0.9

From this distribution, $E(G) = \$0.48$.

Stock four: In this case the distribution of G is

G	$P(G)$
−1.60	0.1
−0.40	0.4
0.80	0.5

For this distribution, $E(G) = \$0.08$.

The choice that maximizes expected gain is to stock three items each day.

1. a. From the distribution of the number of defects per pump given in the problem,

$E(Y) = 2.90$ and $SD(Y) = 2.28$.

The distribution of the mean number of defects per pump in a random sample of 100 will center at $E(Y) = 2.90$ and will have standard deviation given by $= \frac{SD(Y)}{\sqrt{n}} = \frac{2.28}{10} = 0.228$.

b. For the total defects to drop below 230, the mean number of defects per pump must drop below 2.30. This value of 2.30 is about 2.6 standard deviations below the center of the sampling distribution of means. Since the sampling distribution of means should be approximately a normal distribution, a result this small would be very unusual. The facility is not likely to reach this goal of 230 defects unless it improves its basic distribution of defects per pump.

2. a. Turning the frequencies into relative frequencies, or proportions, produces the following:

Y	*P*(*Y*)
16	0.0591
17	0.0885
18	0.1002
19	0.1097
20	0.1140
21	0.1174
22	0.1258
23	0.1400
24	0.1451

b. From the above distribution, $E(Y) = 20.56$ years and $SD(Y) = 2.48$ years. The sampling distribution of the sample mean for a sample of size 400 will be approximately normal in distribution with mean of 20.56 and standard deviation of

$\frac{2.48}{20} = 0.124$.

c. A mean age of 21 is about 3.5 standard deviations above the center of the sampling distribution for means from random samples. Such a result would be very unusual, since the sampling was done randomly. It might be that the sample was not random or that the population being sampled actually included some drivers older than young adult.

3. a. The approximate margin of error is 2 estimated standard deviations of the sample proportion. This turns out to be

$$2SD(\text{proportion}) = 2\sqrt{\frac{(0.81)(0.19)}{1000}} = 0.0248.$$

b. The margin of error is a distance between the sample proportion and the population proportion that will be bettered 95% of the time in repeated use of the technique. With probability 0.95, the sample proportion of those who believe in heaven will not differ from the population proportion who believe the same by more than 0.0248, in samples of size 1000. You can be confident that the 0.81 observed in the sample is reasonably close to the true proportion of those who believe in the existence of heaven.

4. a. The potential values of the sample proportion of those without health insurance, in a random sample of size 500, would have a normal distribution with mean 0.15 and standard deviation given by

$$SD(\text{proportion}) = \sqrt{\frac{(0.15)(0.85)}{500}} = 0.016.$$

b. To find at least 100 Americans with no health insurance in a sample of 500, the sample proportion of those without health insurance would have to be at least 0.2. This value is over three standard deviations above the mean of the sampling distribution. Thus, it is very unlikely that the agency will obtain 100 people without health insurance in a random sample of only 500.

1. a. The probability distribution for a binomial random variable with $n = 6$ and $p = 0.15$ is shown below in a table and in a plot. The distribution is highly skewed toward more patients because the small value of p concentrates most of the outcomes around 0 and 1. Note that 0 does not have the highest probability of occurrence, even though 0.15 is closer to 0 than it is to 1.

The six patients would have to behave as a random selection from a population of patients with 15% having no health insurance in order for these probabilities to be valid. If, for example, the hospital was in a section of a city where more than 15% are without health insurance, then this would not be the correct distribution.

Y	*P(Y)*
0	0.377149
1	0.399335
2	0.176177
3	0.041453
4	0.005486
5	0.000387
6	0.000011

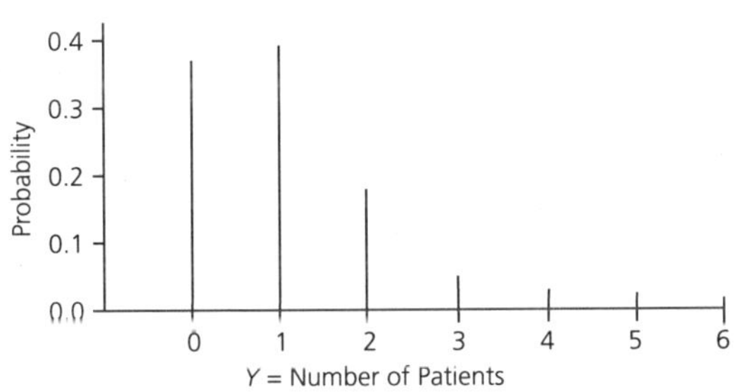

b. For a binomial distribution with $n = 50$ and $p = 0.15$, $E(Y) = np = 50(0.15) = 7.5$ and $SD(Y)$ $= \sqrt{n(p)(1-p)} = \sqrt{50(0.15)(9.85)} = 2.52$.

Again, the sample of 50 must be a random sample from a population with 15% having no health insurance in order for these quantities to be valid.

c. Possible answer: A quick answer is that, since 10 is about 1 standard deviation above the mean, the chance of seeing more than 10 patients without health insurance is about 0.16. The binomial distribution for $n = 50$ will be approximately normal even though p is quite small. The exact answer for the sum of the binomial probabilities greater than 10 is 0.120.

2. a. The measured weight of a battery is out of control if it is more than 3 standard deviations from the mean, under a normal distribution. This probability is 0.003. The chance of not being out of control is 0.997. For independent measurements, the chance of not being out of control 8 times in a row is $(0.997)^8 = 0.976$.

b. If the control limits shift to 2 standard deviations from the mean, then the probability of being within them is 0.95. The probability of being within them 8 times in a row is $(0.95)^8 = 0.663$. Note that this change in the control limits makes a big difference in the chances of being out of control at least once during the day.

3. a. If Y is the number of hours or the number of sampled batteries until an out-of-control signal flashes, then Y has a geometric distribution and $E(Y) = \frac{1}{0.003} = 333.3$.

b. $SD(Y) = \frac{\sqrt{1-p}}{p} = \frac{\sqrt{0.997}}{0.003} = 332.8$

Note that the expected value and standard deviation are nearly equal when p is small.

c. The quick answer is that, since 600 is only about 1 standard deviation above the mean, it must have a good chance of occurring. The exact answer is $(0.997)^{600} = 0.165$. This is most easily calculated by using logarithms.

4. Think of this as a series of three waiting-time problems. When starting out, you will want whichever prize comes up, so the probability of success is 1. At the second stage, you are looking for one of the two remaining prizes, so your probability of success is $\frac{2}{3}$. After you get one of the two, you are looking for the single prize that is not in your collection. Thus, at the third stage your probability of success is $\frac{1}{3}$. Since all the selections of prizes are independent, your expected number of tubs of popcorn is the sum of the expected waiting times for the three stages, namely, $1 + \frac{3}{2} + \frac{3}{1} = 5.5$.

You can expect to buy 5.5 tubs of popcorn to complete your collection.

Data-Driven Mathematics
Procedures for Using the TI-83

I. Clear menus

ENTER will execute any command or selection. Before beginning a new problem, previous instructions or data should be cleared. Press **ENTER** after each step below.

1. To clear the function menu, **Y=**, place the cursor anyplace in each expression, **CLEAR**
2. To clear the list menu, **2nd MEM ClrAllLists**
3. To clear the draw menu, **2nd DRAW ClrDraw**
4. To turn off any statistics plots, **2nd STATPLOT PlotsOff**
5. To remove user created lists from the Editor, **STAT SetUpEditor**

II. Basic information

1. A rule is active if there is a dark rectangle over the option. See Figure 1.

```
Plot1 Plot2 Plot3
\Y1 = 170 + 7(X–72)
\Y2 = –.35X – 22
\Y3 =
\Y4 =
\Y5 =
\Y6 =
\Y7 =
```

FIGURE 1

On the screen above, Y1 and Plot1 are active; Y2 is not. You may toggle Y1 or Y2 from active to inactive by putting the cursor over the = and pressing **ENTER**. Arrow up to **Plot1** and press **ENTER** to turn it off; arrow right to **Plot2** and press **ENTER** to turn it on, etc.

2. The Home Screen (Figure 2) is available when the blinking cursor is on the left as in the diagram below. There may be other writing on the screen. To get to the Home Screen, press **2nd QUIT**. You may also clear the screen completely by pressing **CLEAR**.

```
7 → X
                7
Y1
               11
█
```

FIGURE 2

3. The variable x is accessed by the **X, T, Θ, *n*** key.
4. Replay option: **2nd ENTER** allows you to back up to an earlier command. Repeated use of **2nd ENTER** continues to replay earlier commands.
5. Under **MATH**, the **MATH** menu has options for fractions to decimals and decimals to fractions, for taking *n*th roots, and for other mathematical operations. **NUM** contains the absolute value function as well as other numerical operations. (Figure 3)

```
MATH NUM CPX PRB
1: abs(
2: round(
3: iPart(
4: fPart(
5: int(
6: min(
7↓max(
```

FIGURE 3

III. The STAT Menus

1. There are three basic menus under the **STAT** key: **EDIT**, **CALC**, and **TESTS**. Data are entered and modified in the **EDIT** mode; all numerical calculations are made in the **CALC** mode; statistical tests are run in the **TEST** mode.
2. **Lists and Data Entry**
   Data is entered and stored in Lists (Figure 4). Data will remain in a list until the list is cleared. Data can be cleared using **CLEAR** L_i or (List name), or by placing the cursor over the List heading and selecting **CLEAR ENTER.** To enter data, select **STAT EDIT** and with the arrow keys move the cursor to the list you want to use.

Type in a numerical value and press **ENTER.** Note that the bottom of the screen indicates the List you are in and the list element you have highlighted. 275 is the first entry in L1. (It is sometimes easier to enter a complete list before beginning another.)

L1	L2	L3
275	67	190
5311	144	120
114	64	238
2838	111	153
15	90	179
332	68	207
3828	94	153
L1 (1) = 275		

FIGURE 4

For data with varying frequencies, one list can be used for the data, and a second for the frequency of the data. In Figure 5 below, the L5(7) can be used to indicate that the seventh element in list 5 is 4, and that 25 is a value that occurs 4 times.

L4	L5	L6
55	1	-------
50	3	
45	6	
40	14	
35	12	
30	9	
25	4	
L5 (7) = 4		

FIGURE 5

3. **Naming Lists**
Six lists are supplied to begin with. L1, L2, L3, L4, L5, and L6 can be accessed also as **2nd L_i.** Other lists can be named using words as follows. Put the cursor at the top of one of the lists. Press **2nd INS** and the screen will look like that in Figure 6.

	L1	L2 1
	-------	-------
Name =		

FIGURE 6

The alpha key is on, so type in the name (up to five characters) and press **ENTER.** (Figure 7)

PRICE	L1	L2 2
	-------	-------
PRICE(1) =		

FIGURE 7

Then enter the data as before. (If you do not press **ENTER,** the cursor will remain at the top and the screen will say "error: data type.") The newly named list and the data will remain until you go to Memory and delete the list from the memory. To access the list for later use, press **2nd LIST** and use the arrow key to locate the list you want under the **NAMES** menu. You can accelerate the process by typing **ALPHA P** (for price). (Figure 8) Remember, to delete all but the standard set of lists from the editor, use SetUp Editor from the **STAT** menu.

NAMES OPS MATH
↑PRICE
: RATIO
: RECT
: RED
: RESID
: SATM
↓SATV

FIGURE 8

4. Graphing Statistical Data

General Comments

- Any graphing uses the **GRAPH** key.
- Any function entered in Y1 will be graphed if it is active. The graph will be visible only if a suitable viewing window is selected.
- The appropriate *x*- and *y*-scales can be selected in **WINDOW**. Be sure to select a scale that is suitable to the range of the variables.

Statistical Graphs

To make a statistical plot, select **2nd Y=** for the **STAT PLOT** option. It is possible to make three plots concurrently if the viewing windows are identical. In Figure 9, Plots 2 and 3 are off, Plot 1 is a scatter plot of the data (Costs, Seats), Plot 2 is a scatter plot of (L3,L4), and Plot 3 is a box plot of the data in L3.

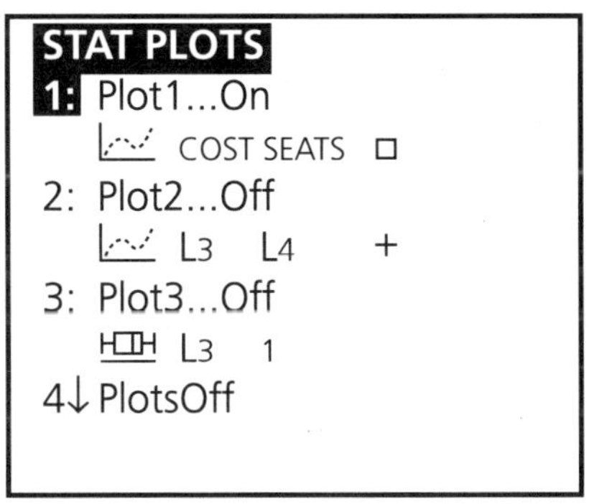

FIGURE 9

Activate one of the plots by selecting that plot and pressing **ENTER.**

Choose **ON,** then use the arrow keys to select the type of plot (scatter, line, histogram, box plot with outliers, box plot, or normal probability plot). (In a line plot, the points are connected by segments in the order in which they are entered. It is best used with data over time.) Choose the lists you wish to use for the plot. In the window below, a scatter plot has been selected with the *x*-coordinate data from COSTS, and the *y*-coordinate data from SEATS. (Figure 10) (When pasting in list names, press **2nd LIST,** press **ENTER** to activate the name, and press **ENTER** again to locate the name in that position.)

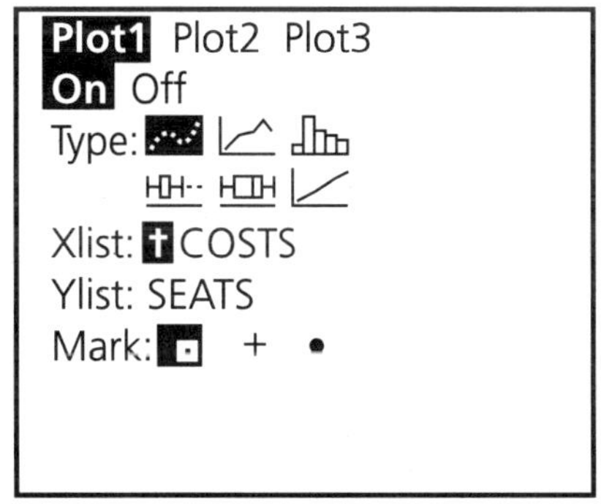

FIGURE 10

For a histogram or box plot, you will need to select the list containing the data and indicate whether you used another list for the frequency or are using 1 for the frequency of each value. The *x*-scale selected under **WINDOW** determines the width of the bars in the histogram. It is important to specify a scale that makes sense with the data being plotted.

5. Statistical Calculations

One-variable calculations such as mean, median, maximum value of the list, standard deviation, and quartiles can be found by selecting **STAT CALC 1-Var Stats** followed by the list in which you are interested. Use the arrow to continue reading the statistics. (Figures 11, 12, 13)

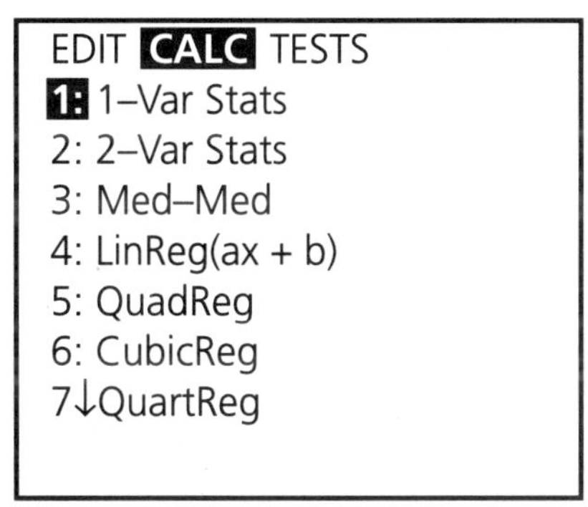

FIGURE 11

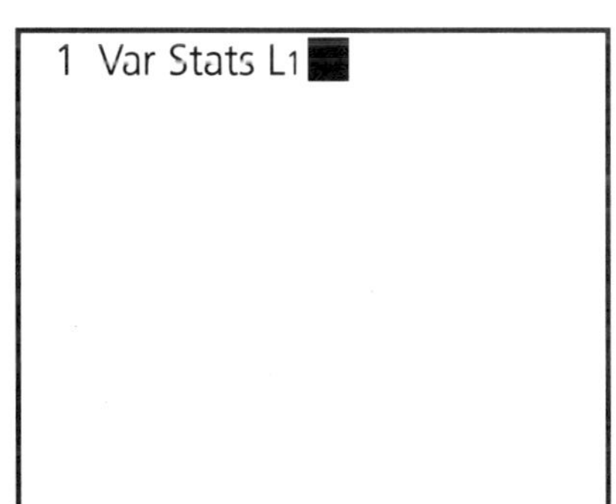

FIGURE 12

```
1–Var Stats
x̄ = 1556.20833
Σx = 37349
Σx² = 135261515
Sx = 1831.353621
σx = 1792.79449
↓n = 24
■
```

FIGURE 13

Calculations of numerical statistics for bivariate data can be made by selecting two variable statistics. Specific lists must be selected after choosing the **2-Var Stats** option. (Figure 14)

```
2–Var Stats L1, L
2■
```

FIGURE 14

Individual statistics for one- or two-data sets can be obtained by selecting **VARS Statistics,** but you must first have calculated either 1-Var or 2-Var Statistics. (Figure 15)

```
XY Σ EQ TEST PTS
1: n
2: x̄
3: Sx
4: σx
5: ȳ
6: Sy
7↓σy
```

FIGURE 15

6. **Fitting Lines and Drawing Their Graphs** Calculations for fitting lines can be made by selecting the appropriate model under **STAT CALC: Med-Med** gives the median fit regression, **LinReg** the least-squares linear regression, and so on. (Note the only difference between **LinReg (ax+b)** and **LinReg (a+bx)** is the assignment of the letters a and b.) Be sure to specify the appropriate lists for x and y. (Figure 16)

```
Med–Med LCal, LFA
CAL■
```

FIGURE 16

To graph a regression line on a scatter plot, follow the steps below:

- Enter your data into the Lists.
- Select an appropriate viewing window and set up the plot of the data as above.
- Select a regression line followed by the lists for x and y, **VARS Y-VARS Function** (Figures 17, 18) and the Y_i you want to use for the equation, followed by **ENTER.**

```
VARS Y-VARS
1: Function...
2: Parametric...
3: Polar...
4: On/Off...
```

FIGURE 17

```
Med–Med _CAL, LFA
CAL, Y1
```

FIGURE 18

The result will be the regression equation pasted into the function Y1. Press **GRAPH** and both the scatter plot and the regression line will appear in the viewing window. (Figures 19, 20)

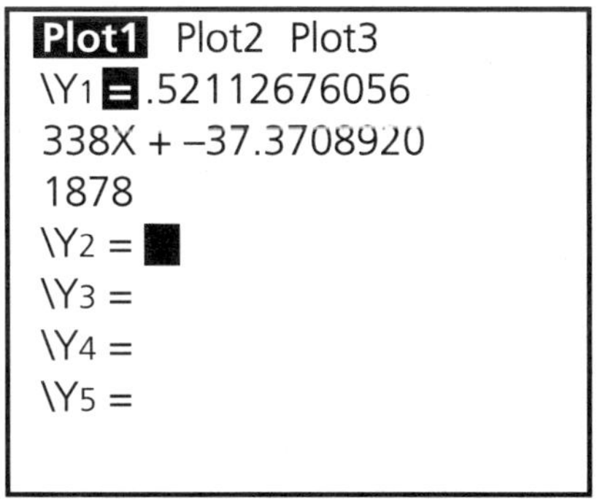

FIGURE 19

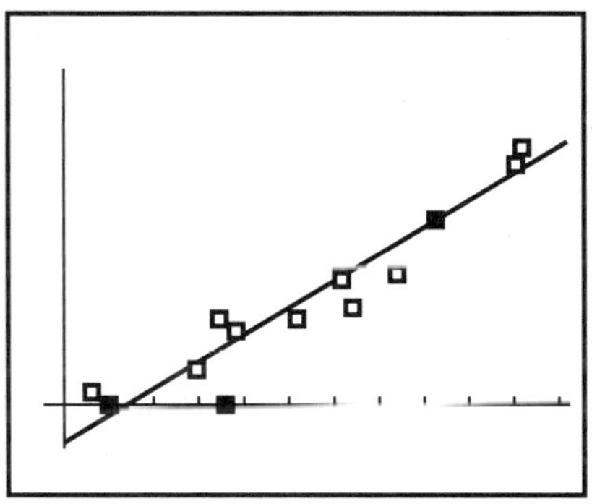

FIGURE 20

- There are two cursors that can be used in the graphing screen.

TRACE activates a cursor that moves along either the data (Figure 21) or the function entered in the Y-variable menu (Figure 22). The coordinates of the point located by the cursor are given at the bottom of the screen.

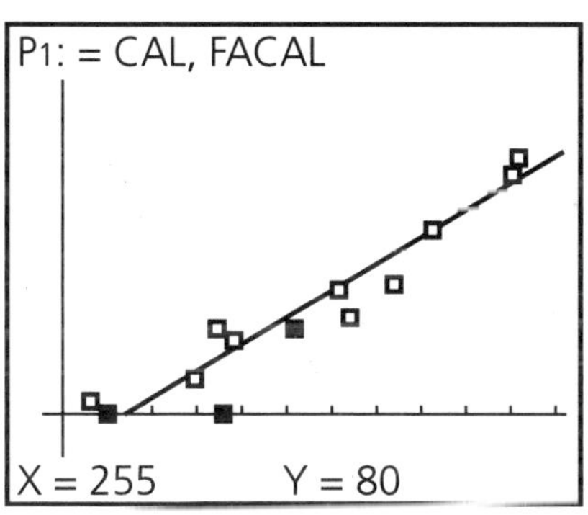

FIGURE 21

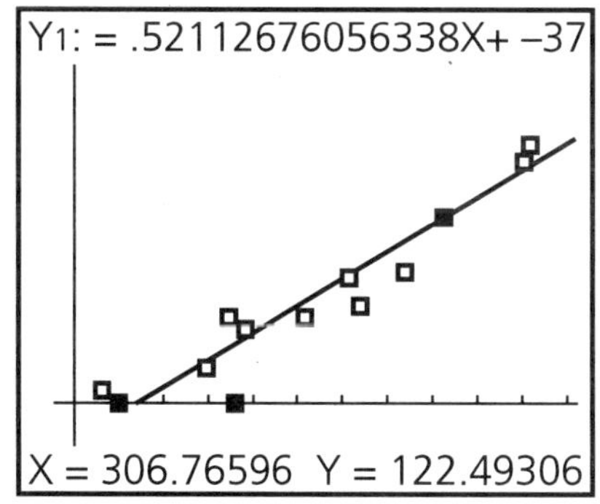

FIGURE 22

Pressing **GRAPH** returns the screen to the original plot. The up arrow key activates a cross cursor that can be moved freely about the screen using the arrow keys. See Figure 23.

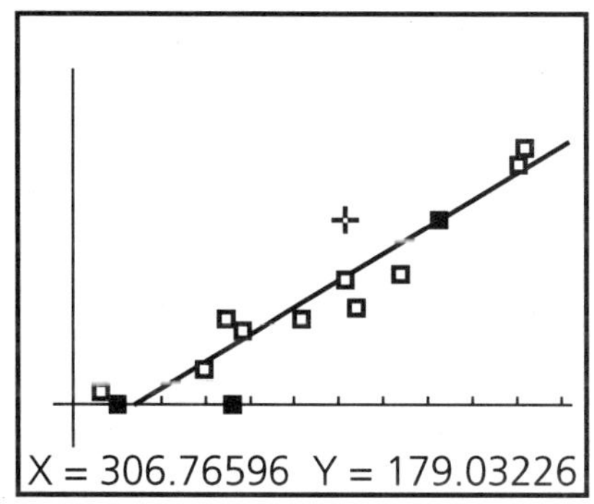

FIGURE 23

Exact values can be obtained from the plot by selecting **2nd CALC** Value. Select **2nd CALC Value ENTER**. Type in the value of x you would like to use, and the exact ordered pair will appear on the screen with the cursor located at that point on the line. (Figure 24)

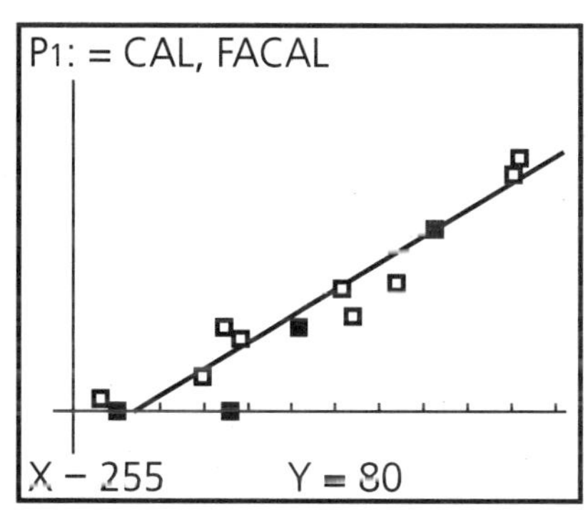

FIGURE 24

IV. Evaluating an expression

To evaluate $y = .225\,x - 15.6$ for $x = 17, 20$, and 24, you can:

1. Type the expression in Y1, return to the home screen, 17 **STO X,T,Θ,n ENTER, VARS Y-VARS Function Y1 ENTER ENTER.** (Figure 25)

```
17 → X
                         17
Y1
             −11.775
■
```

FIGURE 25

Repeat the process for $x = 20$ and 24.

2. Type $17^2 - 4$ for $x = 17$, **ENTER** (Figure 26). Then use **2nd ENTRY** to return to the arithmetic line. Use the arrows to return to the value 17 and type over to enter 20.

```
.225*17 – 15.6
                    −11.775
.225*2█ – 15.6
```

FIGURE 26

You can also find the value of x by using the table command. Select **2nd TblSet** (Figure 27). (Y1 must be turned on.) Let **TBlStart** = 17, and the increment **ΔTbl** = 1.

```
TABLE SETUP
  TblStart = 17
  ΔTbl = █
Indpnt:  Auto  Ask
Depend:  Auto  Ask
```

FIGURE 27

Select **2nd TABLE** and the values of x and y generated by the equation in Y1 will be displayed. (Figure 28)

X	Y1	
17	-11.78	
18	-11.55	
19	-11.33	
20	-11.1	
21	-10.88	
22	-10.65	
23	-10.43	

X = 17

FIGURE 28

V. Operating with Lists

1. A list can be treated as a function and defined by placing the cursor on the label above the list entries. List 2 can be defined as L1 + 5. (Figure 29)

L1	L2	L3
275	-------	190
5311		120
114		238
2838		153
15		179
332		207
3828		153

L2 = L1 + 5

FIGURE 29

Pressing **ENTER** will fill List 2 with the values defined by L1+5. (Figure 30)

L1	L2	L3
275	**280**	190
5311	5316	120
114	119	238
2838	2843	153
15	20	179
332	337	207
3828	3833	153
L2(1) = 280		

FIGURE 30

2. List entries can be cleared by putting the cursor on the heading above the list, and selecting **CLEAR** and **ENTER.**

3. A list can be generated by an equation from Y= over a domain specified by the values in L1 by putting the cursor on the heading above the list entries. Select **VARS Y-VARS Function Y1 ENTER (L1) ENTER.** (Figure 31)

L1	L2	**L3**
120	12	-------
110	14	
110	12	
110	11	
100	?	
100	6	
120	9	
L3 = Y1(L1)		

FIGURE 31

4. The rule for generating a list can be attached to the list and retrieved by using quotation marks (**ALPHA** +) around the rule. (Figure 32) Any change in the rule (Y1 in the illustration) will result in a change in the values for L1. To delete the rule, put the cursor on the heading at the top of the list, press **ENTER**, and then use the delete key. (Because L1 is defined in terms of CAL, if you delete **CAL** without deleting the rule for L1 you will cause an error.)

CAL	FACAL	**L1** 5
255	80	-------
305	120	
410	180	
510	250	
320	90	
370	125	
500	235	
L1 = "Y1(LCAL)"		

FIGURE 32

VI. Using the DRAW Command

To draw line segments, start from the graph of a plot, press **2ND DRAW**, and select **Line(**. (Figure 33)

DRAW POINTS STO
1: ClrDraw
2: Line(
3: Horizontal
4: Vertical
5: Tangent(
6: DrawF
7↓Shade(

FIGURE 33

This will activate a cursor that can be used to mark the beginning and ending of a line segment. Move the cursor to the beginning point and press **ENTER**; use the cursor to mark the end of the segment, and press **ENTER** again. To draw a second segment, repeat the process. (Figure 34)

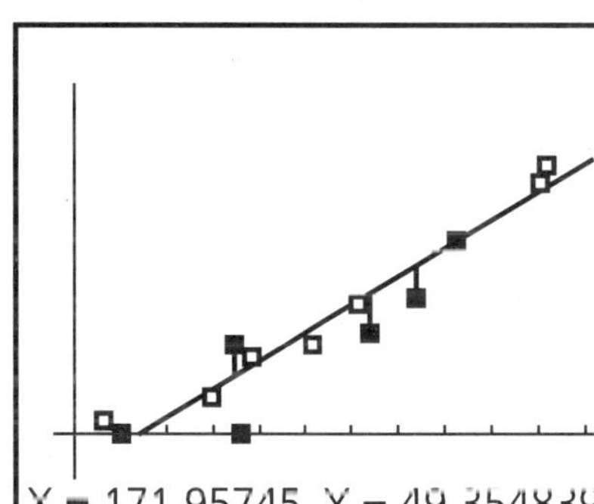

FIGURE 34

VII. Random Numbers

To generate random numbers, press **MATH** and **PRB**. This will give you access to a random number function, **rand**, that will generate random numbers between 0 and 1 or **randInt(** that will generate random numbers from a beginning integer to an ending integer for a specified set of numbers. (Figure 35) In Figure 36, five random numbers from 1 to 6 were generated.

```
MATH  NUM  CPX  PRB
1: rand
2: nPr
3: nCr
4: !
5: randInt(
6: randNorm(
7: randBin(
```

FIGURE 35

```
randInt(1, 6, 5,)
      (2  4  2  5  3)
```

FIGURE 36

Pressing **ENTER** will generate a second set of random numbers.